MOLECULAR ASPECTS OF EARLY DEVELOPMENT

MOLECULAR ASPECTS OF EARLY DEVELOPMENT

**Edited by
George M. Malacinski
and
William H. Klein**

*Indiana University
Bloomington, Indiana*

PLENUM PRESS • NEW YORK AND LONDON

Library of Congress Cataloging in Publication Data

Symposium on Molecular Aspects of Early Development (1982:
  Louisville, Ky.)
  Molecular aspects of early development.

  "Proceedings of the Symposium on Molecular Aspects of Early Devel-
opment, which was part of the annual meeting of the American Society of
Zoologists, held December 29–30, 1982,...Louisville Ky."—Verso t.p.
  Includes bibliographical references and indexes.
  1. Developmental biology—Congresses. 2. Developmental genetics—
Congresses. 3. Molecular biology—Congresses. I. Malacinski, George
M. II. Klein, William H. III. American Society of Zoologists. IV. Title.
  QH491.S95 1982   591.3   83-19259
  ISBN 0-306-41496-1

Proceedings of the Symposium on Molecular Aspects of Early
Development, which was part of the annual meeting of the
American Society of Zoologists, held December 29–30, 1982,
at Galt House, Louisville, Kentucky

© 1984 Plenum Press, New York
A Division of Plenum Publishing Corporation
233 Spring Street, New York, N.Y. 10013

Printed in the United States of America

PREFACE

The early embryo has emerged as the focal point for analysis of the regulation of gene expression for several reasons. First, the fact that embryogenesis is under genetic control has been appreciated from the earliest days of classical embryology. When experimental techniques became available it was therefore logical that they should be applied to the embryo. With each new advance in methodology, interest in embryonic gene expression studies has increased. Second, many embryos offer unique opportunities for the investigation of specific aspects of the regulation of gene expression. Several phenomena--eg., control of translation--can be very conveniently studied in a variety of marine invertebrate embryos. Those embryos contain large stores of maternally inherited mRNA which are translated in a highly ordered fashion during specific stages of post fertilization development. Marine invertebrate eggs can be conveniently artificially inseminated and labeled with radioactive precursors. Their analysis is leading to important insights into the mechanisms which regulate gene expression at post-transcriptional levels.

Third, recent advances in both transmission and recombinant DNA genetics, especially in organisms such as <u>Drosophila</u>, are providing special opportunities for the analysis of regulatory mechanisms which operate at the level of the genome. Specific genes have been identified, isolated, and--in some instances--sequenced. The opportunity is now available to study the regulation of the expression of single genes in a vertical fashion--from the primary sequence of the gene to the tissues and organs which are the products of morphogenesis.

Fourth, studies on the regulation of gene expression in vertebrate embryos promise to provide fundamental information concerning the basis of congenital defects. Consequently, various amphibian and mammalian embryos have been defined as model systems for investigating molecular aspects of early pattern specification.

v

Most embryos employ two gene expression strategies in their
early development: the utilization of maternally inherited
components, including structural proteins, enzymes, ribosomal,
transfer, and messenger RNA; and the de novo synthesis (directed by
the zygote's genome) of novel proteins and nucleic acids. A common
goal of contemporary molecular embryologists is to understand the
manner in which maternal vs. zygote contributions regulate early
pattern specification. Most investigations begin with descriptive
studies which try to determine the extents to which various gene
expression patterns change during early embryogenesis. Those studies
are then usually follwed by analyses of the relative contributions of
the maternal and zygotic genomes. Finally, attempts at identifying
and isolating specific genes usually ensue.

The reports included in this volume represent research progress
towards that common goal. They were presented at a symposium
entitled "Molecular Aspects of Early Development" as part of the
annual meeting of the American Society of Zoologists (Louisville,
Kentucky, December 27-30, 1982). The reports describe how the
embryos from various organisms (nematodes, marine invertebrates,
insects, amphibia, and mammals) are currently being exploited with
modern techniques (protein fractionation, recombinant DNA probes
monoclonal antibodies, in situ hybridization, etc.) to formulate
coherent views of the manner in which gene expression is regulated in
the early embryo.

Symposium Organizers   George M. Malacinski and
                                            William H. Klein

CONTENTS

THE YELLOW CRESCENT OF ASCIDIAN EGGS: MOLECULAR

ORGANIZATION, LOCALIZATION AND ROLE IN EARLY DEVELOPMENT

William R. Jeffery, Craig R. Tomlinson,
Richard D. Brodeur, and Stephen Meier[*]

ABSTRACT

The molecular composition, localization, and role in early
development of the yellow crescent cytoplasm is reviewed. The
yellow, myoplasmic crescent is a localized cytoplasmic region
preferentially distributed to the muscle and mesenchyme lineage cells
during early development of ascidian eggs. It consists of a
collection of lipid pigment granules with numerous adherent
mitochondria underlain by a specific cytoskeletal domain. The yellow
crescent cytoskeleton is comprised of a superficial, sub-membrane
network of actin filaments (PML) and a more internal filamentous
lattice which connects pigment granules and possibly other
cytoplasmic organelles to the cell surface. The yellow crescent
originates during oogenesis and is uniformly distributed around the
periphery of the mature, unfertilized egg. After fertilization the
peripheral cytoplasm streams into the vegetal hemisphere forming the
yellow crescent. The yellow crescent cytoskeleton, under the
direction of local changes in the concentration of calcium ions,
seems to be involved in this movement. Although relatively poor in
total mRNA, the yellow crescent is highly enriched in mRNA sequences
coding for cytoplasmic actin. The enrichment in actin mRNA is due to
an association of these molecules with yellow crescent cytoskeletal
elements. In general, however, prevalent messages in the yellow
crescent region are not qualitatively different from those in other
areas of the egg. A wide variety of different proteins are also
found in the yellow crescent which are a subset of those present in

[*]From the Department of Zoology, University of Texas, Austin,
TX 78712.

the whole egg. There is strong evidence that the yellow crescent contains cytoplasmic determinants which are segregated during cleavage and specify muscle cell properties in the cells they enter. The molecular nature and mode of action of these agents, however, remains to be determined.

INTRODUCTION

Classical experiments conducted in the early part of the century established an important role for specialized regions of the egg cytoplasm in the control of embryonic determination (for reviews see Wilson, 1925; Davidson, 1976). It has been hypothesized that such regions contain morphogenetic determinants, substances segregated to different cell lineages that are responsible for mediating the developmental choices made by totipotent nuclear genomes (Morgan, 1934). Although these determinative agents remain to be isolated and characterized, there is evidence that some of them may be maternal RNA sequences (Kalthoff, 1979; see Jeffery, 1983b, for review) or proteins (Brothers, 1979).

The absence of visible markers in the specialized cytoplasmic regions of eggs has contributed to the difficulty of assessing their developmental significance. There are exceptions to this generalization, however, and some of these have provided our richest sources of information on the role of cytoplasmic factors in early embryogenesis. The oosome, a specialized cytoplasmic region involved in germ cell determination (Illmensee and Mahowald, 1974), is positioned at the posterior pole of dipteran eggs. The oosome contains a unique localization of organelles known as polar granules (see Mahowald et al., 1979, for review). Organelles of similar structure are found in specialized regions of the cytoplasm thought to be involved in germ cell determination in a number of different kinds of eggs (see Jeffery, 1983b, for review).

Cytoplasmic regions of specific morphogenetic fate can also be marked by distinct colors in certain eggs, most notably those of ctenophores (Spek, 1926) and ascidians (reviewed by Berrill, 1968). The greatest variety of different colored cytoplasmic regions was discovered in the egg of the ascidian Styela by Conklin (1905a). One of these regions, the yellow crescent, is the subject of this review. We begin by considering the structure and developmental history of this specialized cytoplasmic region which is a marker for the larval muscle and mesenchyme cell lineages in ascidian eggs. Next, we review classical and more recent studies on the role of the yellow crescent in muscle cell determination and the nature of the determinative agents. Finally, we outline recent studies on the macromolecular organization and mechanism of localization of the yellow crescent region during early development.

Structure of the Yellow Crescent

Many ascidian eggs are characterized by brightly colored cytoplasms, but only a few genera exhibit colored crescents that delimit the territory presumptive for the mesodermal cell lineages. Among these are the yellow crescent of Styela (Conklin, 1905a) and the orange crescent of Boltenia (Berrill, 1929) eggs. The crescents are formed during the period between the end of ooplasmic segregation and the first cleavage (Fig. 1) by the spreading of cortical pigmentation over the posterior side of the vegetal hemisphere of the egg. It has been known since the early part of the century that the crescents are primarily composed of mitochondria and pigment granules. The lipid nature of the pigment particles was first suggested by Conklin (1905a) and later confirmed by their displacement to the most centripetal end of the egg by centrifugation (Conklin, 1931). The abundance of mitochondria in the crescent region has been demonstrated by microchemical analyses of isolated posterior blastomeres (Berg, 1956, 1957), cytochemistry (Reis, 1937; Reverberi and Pitotti, 1939), and vital staining with Janus green (Reverberi, 1956), acridine orange (Zalokar, 1980), and rhodamine 123 (DeSantis and Stopak, 1980).

Structural analysis of the yellow crescent region of Styela eggs has been achieved by electron microscopy (Berg and Humphreys, 1960). This approach provided direct evidence that the crescent is composed of lipid particles with adherent mitochondria (Fig. 2). The crescent region contains about a third of the total egg mitochondria and almost 90% of the yellow pigment granules (Berg and Humphreys, 1960). A similar concentration of mitochondria and lipid granules has been observed in ascidian species with orange (Boltenia; Jeffery 1982a) or colorless (Ciona; Berg and Humphreys, 1960) myoplasmic crescents. Aside from the aggregations of mitochondria and pigment granules, the yellow crescent also seems to be enriched in vesicles of endoplasmic reticulum and small spherical bodies. The latter have been variously described as ribosomes (Berg and Humphreys, 1960; Mancuso, 1963) or glycogen particles (Ursprung and Schabtach, 1964).

Jeffery and Meier (1983) have recently extracted Styela and Boltenia eggs with the non-ionic detergent Triton X-100 and searched for cytoskeletal structures. Transmission electron microscopy revealed an anastomosing system of filamentous structures between the mitochondria, lipid pigment granules, and ribosome-like bodies. Conklin (1905a) noted that the yellow crescent exhibited a granular or alveolar appearance after the pigment granules were displaced by centrifugation. Although this observation has usually been attributed to the retention of mitochondria, assuming they are not displaced by low centrifugal forces (Berg and Humphreys, 1960), it may also be explained by the presence of a rigid cytoskeletal framework.

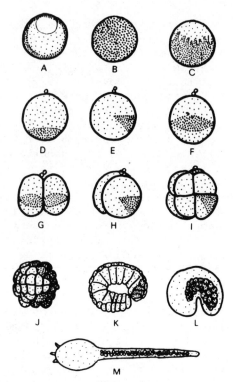

Figure 1:   Developmental history of the yellow crescent cytoplasm of
Styela eggs.   A) A cross-section through the animal-vegetal axis of
a mature oocyte showing the large germinal vesicle and the yellow
cytoplasm in the cell periphery.   B) An unfertilized egg showing
the yellow cytoplasm present uniformly around the cell periphery.
C) A fertilized egg involved in ooplasmic segregation showing the
yellow cytoplasm streaming into the vegetal hemisphere.   D) A
fertilized egg in which the yellow cytoplasm has capped at the
vegetal hemisphere.   E-F) Fertilized eggs in which a yellow
crescent has formed in the posterior vegetal region.   G-H) Two-cell
embryos in which yellow crescent material has been distributed to
both blastomeres.   I) An eight-cell embryo in which yellow crescent
material has been distributed to the two posterior vegetal
blastomeres.   J) A 64-cell embryo in which yellow crescent
cytoplasm has been distributed to the presumptive muscle and
mesenchyme lineage cells.   K) A gastrula showing the position of
the muscle cells.   L) A tail-bud embryo showing muscle cells
beginning to enter the tail.   M) A tadpole larva in which muscle
cells line the tail in parallel rows.

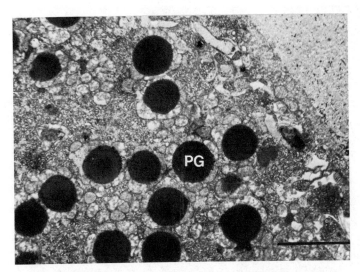

Figure 2: An electron micrograph of the yellow crescent region of
the Styela plicata egg. The yellow crescent contains endoplasmic
reticulum, lipid pigment granules (PG), and adherent mitochondria.
The bar represents 10 microns.

The superficial features of the yellow crescent have been
examined in detergent extracted Styela eggs by scanning electron
microscopy (Jeffery and Meier, 1983). The residue left in the
crescent region appears to be comprised of at least three elements, a
superficial network of filaments (the plasma membrane lamina; PML), a
deeper filamentous lattice, and pigment granules that are embedded in
the lattice. The surface of the crescent cytoskeletal domain is
shown in Fig. 3. The PML is seen as a network of filaments organized
parallel to the original plane of the plasma membrane and restricted
to the outer boundary of the crescent region. This structure is
thought to be a remnant of integral membrane proteins and associated
cytoskeletal elements located immediately below the plasma membrane.
Sawada and Osanai (1981), have observed the possible counterpart of
the PML in the area directly below the colorless myoplasmic crescent
of Ciona eggs by transmission electron microscopy (Fig. 4). Membrane
lamina of this nature have also been reported at the outer boundary
of the cytoskeleton in erythrocytes and other vertebrate somatic
cells (Hainfield and Steck, 1977; Boyles and Bainton, 1979).

The PML appears to be composed in part of actin filaments. This
is supported by two different experiments reported by Jeffery and
Meier (1983). First, as in other eukaryotic cells (Hitchcock et al.,
1976; Raju et al., 1978), actin is selectively removed from the
cytoskeleton of ascidian eggs by DNase I treatment. When eggs
treated with Triton X-100 in the presence of DNase I were examined

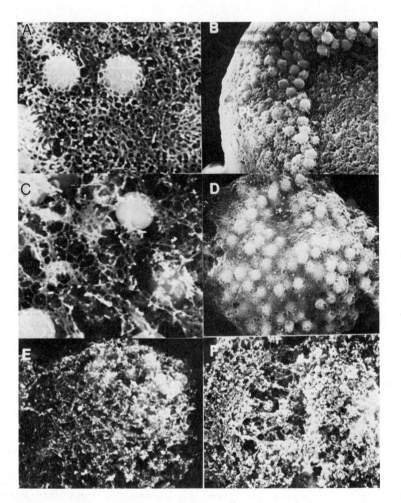

Figure 3: Scanning electron micrographs of the plasma membrane
lamina (PML) of Triton X-100 extracted ascidian eggs. A) The
surface of the yellow crescent region of a fertilized Styela
plicata egg. X 7,500. B) The orange crescent region of a
fertilized Boltenia villosa egg undergoing ooplasmic segregation.
The migration of pigment granules is proceeding to the upper left.
X 150. C) The surface of an unfertilized Styela egg. X 7,500.
Note the greater distance between the filaments of the PML in C as
compared to A. Cytoskeletal frameworks of mesodermal (D),
ectodermal (E), and endodermal (F), lineage cells from a 32-cell
Styela embryo. X 1,900. Only the mesodermal cells contain the
PML. From Jeffery and Meier (1983).

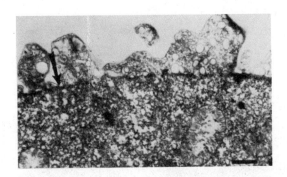

Figure 4: An electron micrograph of the cortical region of the
  vegetal myoplasmic map of a fertilized Ciona intestinalis egg which
  has recently completed ooplasmic segregation. The arrow represents
  the sub-membrane electron dense layer thought to represent the
  plasma membrane lamina. Note also the cytoplasmic blebs outside of
  this layer. The bar represents 1 micron. From Sawada and Osanai
  (1981).

the PML was found to be almost entirely absent. Second, the
peripheral areas of the yellow crescent stain intensely after
treatment with phallacidin, a specific fluorescent indicator of
F-actin (Barak et al., 1980), (Fig. 5). The presence of actin
filaments in the PML suggests that its contraction may be
instrumental in yellow crescent formation.

    The origin and fate of the PML during early development is
closely linked to the history of the yellow crescent cytoplasm (see
below). It is initially located over the entire surface of the
detergent-extracted egg. After fertilization, it appears to recede
into the vegetal hemisphere along with the yellow cytoplasm during
ooplasmic segregation (Fig. 3B). The average distance between the
interstices of the filamentous network are also considerably reduced
after ooplasmic segregation, as if the entire PML contracted toward a
single position on the cell surface (compare Fig. 3A and 3C). The
PML is specifically partitioned to the presumptive mesenchyme and
muscle cells during early embryogenesis. This was demonstrated by
extracting different cell types isolated from 32-cell embryos with
Triton X-100 (Jeffery and Meier, 1983). As illustrated in Fig.
3D-F, scanning electron microscopy of the detergent-extracted cells
indicates that the PML is restricted to cells containing the crescent
material.

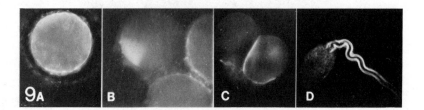

Figure 5: Light micrographs of phallacidin-stained Styela plicata
eggs and embryos. A) An unfertilized egg focused at the edge
showing peripheral actin staining. B) A fertilized egg focused
over part of the yellow crescent showing actin staining the the
crescent region. C) A cleaving egg showing actin staining in the
region of the contractile ring. D) A tadpole larva showing actin
staining of the myofibril bundles in the tail muscle cells. X 100.
In part from Jeffery and Meier (1983).

Another filamentous system is present in the yellow crescent
region beneath the PML. It consists of a filamentous lattice coursed
with bundles of filaments and appears to be attached to both the PML
and the surface of the pigment granules (Fig. 6). The intimate
association with the filamentous lattice probably accounts for the
resistance of the pigment granules to Triton X-100 extraction. The
filamentous lattice may be responsible for the granular or alveolar
appearance of the yellow pigment cytoplasm in light and electron
micrographs.

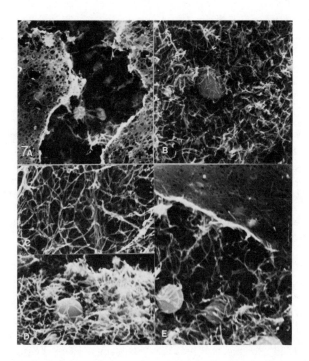

Figure 6: Scanning electron micrographs of cytoskeletal filaments in
   the yellow crescent region of Styela eggs extracted with Triton
   X-100. A) A lesion in the plasma membrane lamina (PML) reveals an
   underlying filamentous lattice. X 4,500. B) A pigment granule
   embedded in the lattice. X 5,000. C) Bundles of filaments often
   course the filamentous lattice. X 5,000. D) An inverted PML
   showing a pigment granule enmeshed in the lattice. X 5,000. E) A
   tear in the PML shows that it is connected to the underlying
   filamentous lattice. X 5,000. From Jeffery and Meier (1983) and
   Jeffery (1983b).

Origin and Fate of the Yellow Crescent

The yellow cytoplasm is originally restricted to the periphery
of mature Styela oocytes (Fig. 1A). The oogenetic origin of this
peripheral yellow cytoplasm is not well understood. Harvey (1927)
suggested that the lipid granules of Ciona oocytes originate in the
test cells, accessory cells that embed themselves in the oocyte

cortex during the later stages of oogenesis. Consistent with this
idea, Kessel and Beams (1965), noted that the test cells of Styela
oocytes synthesize and accumulate yellow pigment granules of the same
appearance as those which are present in the crescent. In contrast,
we have recently observed an abundance of yellow pigment granules
located near the edge of the germinal vesicle (GV), of young Styela
oocytes (Jeffery, unpublished). A number of years ago, Hsu (1962;
1963), also noted clusters of mitochondria, Golgi complexes, and
dense granular organelles known as nuclear emission masses localized
near the GV in pre-vitellogenic Boltenia oocytes (Fig. 7). Although
further studies are clearly warranted, the best developmental
sequence for the origin of the yellow cytoplasm seems to be that it
begins near the GV during vitellogenesis and participates in a
concerted movement to the cortex. It is possible, of course, that
the yellow crescent cytoplasm arises from two different sources, the
perinuclear cytoplasm and the test cells.

After fertilization, the localized cytoplasmic regions of
ascidian eggs are rearranged by a spectacular episode of cytoplasmic
streaming known as ooplasmic segregation. Styela eggs contain three
cytoplasmic regions, the ectoplasm, endoplasm, and yellow crescent
cytoplasm, which occupy well-defined territories (Fig. 1B-C), in the
unfertilized egg (Conklin, 1905b). As mentioned earlier, the yellow
cytoplasm is limited to the cell periphery. The transparent
ectoplasm, which originates primarily from the GV plasm, is located
in the animal hemisphere. The yolky endoplasm fills the vegetal
hemisphere and surrounds the ectoplasm in the animal hemisphere.
Ooplasmic segregation begins during the first maturation division
with the streaming of the yellow cytoplasm into the vegetal
hemisphere where it collects as a deeply pigmented cap (Fig. 1B-D).
It is possible that this cap is focused around the sperm, which
enters the egg in the vegetal region (Conklin, 1905a), but this
remains to be rigorously demonstrated. The yellow cytoplasm is
followed into the vegetal hemsiphere by the ectoplasm, which comes to
rest and then spreads out above the yellow cap at the vegetal pole.
As these cytoplasms move through the egg, the endoplasm is displaced
upwards into the animal hemisphere, and protoplasmic lobes are
transiently protruded from the polar regions, first in the animal
hemsiphere and then in the vegetal hemisphere (Fig. 8), (Zalokar,
1974; Sawada and Osanai, 1981). When the second maturation division
begins, the yellow cytoplasm and ectoplasm migrate up along the
vegetal cortex to the future posterior region of the embryo where
they are both extended into crescent-shaped patches (Fig. 1E-F).
Ooplasmic segregation is completed after the ectoplasm, accompanied
by the male pronucleus, streams back into the animal hemisphere, and
the endoplasm returns to its original position in the vegetal
hemisphere.

The developmental fate of the yellow crescent is well known due
to careful cell-lineage studies conducted by Conklin (1905a),

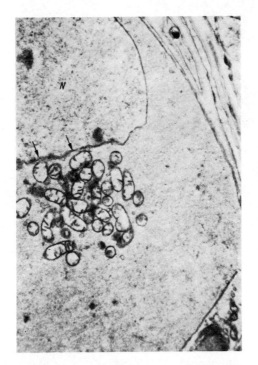

Figure 7: An electron micrograph of a portion of a very young
  Boltenia villosa oocyte showing the mass of mitochondria and
  associated granules near the edge of the germinal vesicle(N).
  X 9,000. From Hsu (1963).

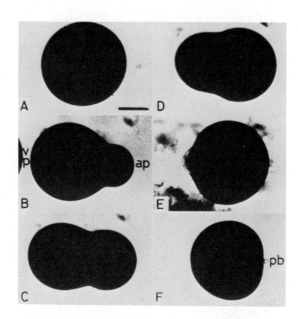

Figure 8: Light micrographs showing the modification of egg shape during the first maturation division of dechorionated Ciona intestinalis eggs. A) An unfertilized egg (primary oocyte). B-F) Successive shape modifications which occur after fertilization as the myoplasmic crescent is formed at the vegetal pole. ap, animal pole; vp, vegetal pole; pb, polar body. The bar in A represents 50 microns. From Sawada and Osanai (1981).

Ortolani (1955a), and others. The first cleavage is vertical and bisects the crescent distributing yellow myoplasm to both daughter cells (Fig. 1G-H). As a result of the second cleavage, which is vertical and at right angles to the first cleavage, the crescent material is partitioned to only two of the four blastomeres. During subsequent cleavages, the yellow crescent cytoplasm is progressively restricted to the larval muscle and mesenchyme cells. At the eight-cell stage, it is found in two cells (Fig. 1I), at the 16-cell stage in four cells and at a 32-cell stage in six cells. The six yellow cells of the 32-cell embryo constitute the stem line for the mesoderm. They cleave into 12 daughter cells; eight of these will become mesenchyme cells while the remaining four cells will become tail muscle cells. The muscle cells divide and are arranged into two parallel rows on each side of the embryo after gastrulation (Fig. 1J-M). Myofibrils differentiate within their cytoplasm as the larval tail is extended.

Morphogenetic Role of the Yellow Crescent

The segregation of yellow cytoplasm to muscle and mesenchyme cell lineages suggested by Conklin (1905a), predicts that the determinants for these tissues may be present in the myoplasmic crescent. Blastomere deletion experiments carried out by Conklin (1905c; 1906) and others and elegant cytoplasmic transfer experiments recently conducted by Whittaker (1980; 1982) support this hypothesis.

The presence of a tough chorion, at first only removable by tedious surgical means, made routine blastomere deletion experiments with developing Styela eggs very difficult. Consequently, Conklin (1905c; 1906), developed alternate methods to study the developmental potential of the cells containing the various egg cytoplasmic regions. In one of his methods, chorionated two-, four-, and eight-cell embryos were vigorously shaken in vials, resulting in the injury and developmental arrest of particular cells. When the partial embryos were cultured each portion gave rise only to those structures it was destined to form in the whole embryo. For example, anterior portions that contained living cells lacking the yellow crescent formed embryos expressing normal ectodermal and endodermal derivatives but no muscle or mesenchyme lineage cells. On the other hand, posterior portions that contained cells with crescent material formed the mesodermal lineage cells but did not express certain ectodermal derivatives.

The problem of the chorion was eventually overcome by Berrill (1932), who used proteolytic enzymes to affect its removal and was able to study the development of isolated blastomeres. His experiments, and those of many others that followed (Tung, 1934; von Ubisch, 1940; Reverberi and Minganti, 1946; Pisano, 1949), suggest that cells containing crescent cytoplasm must be present to promote the development of normal embryonic muscle and mesenchyme lineage cells.

More direct evidence for the role of the yellow crescent in muscle cell determination has recently been reported by Whittaker (1980; 1982). He used two different approaches to transfer cytoplasm from blastomeres bearing the myoplasmic crescent into cells of other morphogenetic assignments. First, he altered the normal segregation of myoplasm at the third cleavage of Styela embryos by compression so that four rather than two blastomeres of the eight-cell stage received yellow crescent material. Second, he was able to transfer myoplasm into presumptive epidermal lineage cells of Ascidia embryos during the third cleavage by a clever microsurical alteration of the position of the cleavage furrow. In both of these experiments he followed subsequent muscle cell determination by the histochemical development of acetylcholinesterase, an enzyme expressed specifically by the developing muscle cells (Durante, 1956; Whittaker, 1973). The

blastomeres that received myoplasm in both types of experiments
eventually developed acetylcholinesterase, an indication that
specific cytoplasmic determinants were transferred to them from cells
containing the myoplasmic crescent.

Although the cytoplasmic transfer experiments indicate that
yellow crescent material (or the cytoplasm surrounding it), can
promote acetylcholinesterase development, the presence of this
cytoplasm does not appear to be sufficient for myofibril
differentiation. Pucci-Minafra and Ortolani (1968), isolated the two
posterior vegetal blastomeres of eight-cell Ciona embryos, cultured
them, and checked for myofibrils by electron microscopy. Bundles of
myofibrils were not distinguished, although the cells underwent three
to four divisions in culture. Myofibril bundles did appear in
presumptive muscle cells, however, when cells from the epidermal
lineage were added to the cultures. Although there are a number of
possible interpretations for this result, the simplest conclusion
would be that yellow crescent material and further morphogenetic
events, possibly involving the interaction of the presumptive muscle
and epidermal cells (Pucci-Minafra and Ortolani, 1968), are required
for myofibril differentiation.

Nature of the Yellow Crescent Determinants

The experiments discussed above suggest that cytoplasmic agents
specifying muscle and mesenchyme cell development are present in the
yellow crescent region of ascidian eggs. Although their identity is
still unknown, it is unlikely that these morphogenetic determinants
include the pigment granules or mitochondria. Conklin (1931),
performed low speed centrifugation experiments in which the yellow
pigment granules (and the mitochondria?) were centrifuged to
atypical locations in the egg just prior to the first cleavage. The
centrifuged particles were unable to return to their normal positions
before cleavage was completed and were deposited in only one of the
two daughter cells. Nevertheless, normal muscle cells developed in
many of these cases. Conversely, mitochondria displaced to atypical
regions of the embryo by centrifugation do not induce muscle cell
development in these areas (Tung et al., 1941).

Another observation also suggests that mitochondria are not
causally involved in muscle cell determination. Some species of
Molgula exhibit anuran larvae. Although their eggs do not show the
mitochondrial localization of their counterparts in urodele species,
the development of vestigial tail muscle acetylcholinesterase
normally occurs during embryogenesis (Whittaker, 1979).

Jeffery and Meier (unpublished), have recently repeated
Conklin's centrifugation experiments and examined the nature of the
yellow crescent region after Triton X-100 extraction. As shown in
Fig. 9A, the filamentous network of the PML appears to be intact

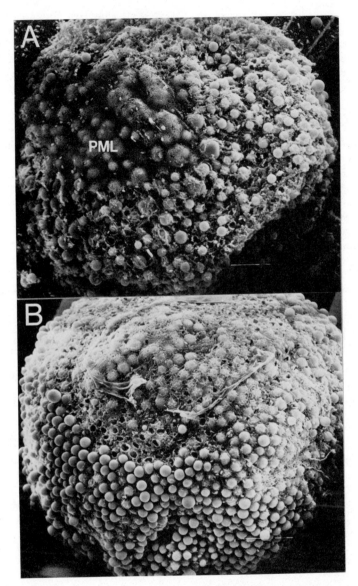

Figure 9: Scanning electron micrographs of centrifuged Styela plicata zygotes extracted with Triton X-100. A) A zygote centrifuged at 200 x g for 2 min. The plasma membrane lamina (PML) is not displaced, but underlying pigment granules are dispersed toward the centripetal end of the egg. B) A zygote centrifuged at 1200 x g for 5 min. The pigment granules are displaced to the centripetal end of the egg, and the PML is apparent but dispersed relative to A. Both frames, X 890.

after low speed centrifugation, although the pigment granules formerly embedded in this region are displaced. Centrifugation at speeds which completely stratify the contents of the egg, and also interfere with normal muscle cell development (Conklin, 1931), tend to disperse the PML (Fig. 9B). These results suggest that a centrifugation-resistant cytoskeletal framework may exist in the crescent region which could prevent low centrifugal forces from displacing the muscle cell determinants.

Macromolecules, such as maternal RNA sequences and proteins, are currently the most popular candidates for egg cytoplasmic determinants. There is some evidence that maternal mRNA, or factors controlling the translation of maternal mRNA, is the determinant for alkaline phosphatase development in the endodermal lineage cells of Ciona embryos (Whittaker, 1977). This does not appear to be the case for the development of acetylcholinesterase in the muscle lineage cells because suppression can be achieved by culturing embryos in actinomycin D (Whittaker, 1973). This result does not exclude the possibility that some of the other characteristics of determined muscle cells are specified by mRNA determinants, however.

Three lines of investigation are now in progress which may improve our understanding of the molecular nature of the cytoplasmic determinants in the yellow crescent. First, microinjection assays are being developed (Whittaker, 1982), which, when applied with the proper enzymatic digestions or other inactivating treatments, may eventually permit the identification of the muscle acetylcholinesterase determinants. Second, UV irradiation and photoreactivation studies, along the lines of the experiments recently conducted with the anterior determinants of Smittia eggs (see Kalthoff, 1979 for review), are being carried out on the yellow crescent cytoplasm. Possibly, this approach will provide information on whether muscle cell determinants are nucleic acids or proteins. Third, characterization of the macromolecular constituents of the yellow crescent is underway using cytological and biochemical methods. The latter studies, which have begun to yield information about the different mRNA sequences and proteins present in the crescent, will be considered in detail in the subsequent sections of this review.

Messenger RNA Sequences of the Yellow Crescent

The nature of yellow crescent mRNA has been studied by Jeffery and his collaborators (Jeffery and Capco, 1978; Jeffery et al., 1983), using in situ hybridization. This approach permits the distribution of total mRNA and individual mRNA species to be mapped with a great degree of precision because different cytoplasmic regions can be readily distinguished in stained sections of fixed Styela eggs and embryos (Fig. 10)

Figure 10: A light micrograph stained with Harris hematoxylin-eosin
to show the three different cytoplasmic regions of fertilized
Styela plicata eggs: the yellow crescent (YC; granular lower
portion), the ectoplasm (dark-staining central portion), and the
endoplasm (granular upper portion). X 400. From Jeffery et al.
(1983). AP, animal pole; VP, vegetal pole.

     The distribution of total mRNA was initially investigated by in
situ hybridization with a radioactively labeled poly(U) probe
(Jeffery and Capco, 1978). These studies showed that the ectoplasm
was labeled eight fold higher than the endoplasm at all stages of
early development (Fig. 11). When these data were quantified, almost
half of the total poly(A) was estimated to be present in the
ectoplasm, a region whose volume represents only 11% of the egg.
Unfortunately, poor fixation in the early experiments prevented a
precise determination of the poly(A) titer of the yellow crescent.
Fixation techniques which preserve the pigment granules have now been
developed and an accurate determination has been made of the extent
of labeling in the yellow crescent (Fig. 11). The yellow crescent
cytoplasm was found to be poorly labeled at each stage of
development. This suggested that the crescent region may be
deficient in mRNA, but further study was necessary because it was
possible that crescent mRNA lacked poly(A) sequences or contained
poly(A) sequences too short to form stable hybrids with the poly(U)
probe. To resolve this issue, Jeffery et al. (1983) prepared
radioactively labeled cDNA from total egg poly(A)$^+$ RNA and used these
sequences as a probe for mRNA. The results were similar to those
obtained with the poly(U) probe (Fig. 12). Most of the grains were
concentrated in the ectoplasm and some were present in the endoplasm,
but only low levels of labeling appeared in the yellow crescent. The
poly(U) and cDNA in situ hybridization results suggest that the
yellow crescent region is poor in mRNA. One possibility for the
paucity of mRNA may be that the yellow crescent is deficient in

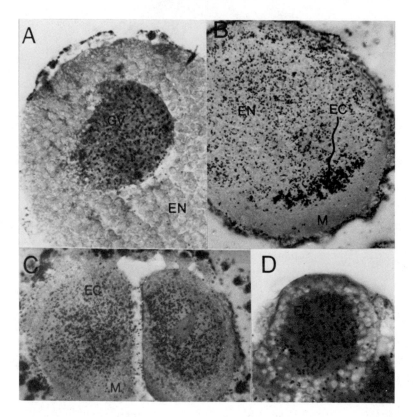

Figure 11: In situ hybridization of sections of Styela plicata eggs
and embryos with a [^{3}H]-poly(U) probe. A) A post-vitellogenic
oocyte with focus on grains concentrated in the germinal vesicle
(GV). B) A fertilized egg which has completed the first part of
ooplasmic segregation. The cytoplasmic regions are stratified
along the animal-vegetal axis in the sequence
endoplasm-ectoplasm-yellow crescent. Many grains are concentrated
in the ectoplasm (EC). Some grains are also present in the
endoplasm (EN), but only a few grains are seen in the yellow
crescent (M). C) A two-cell embryo with many grains present in the
ectoplasm (EC), but few grains in the yellow crescent cytoplasm
(M). D) An eight-cell embryo showing grains concentrated in the
ectoplasm (EC) of an animal hemisphere blastomere. All frames are
X 400. From Jeffery and Capco (1978) and Jeffery et al. (1983).

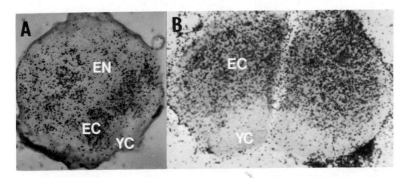

Figure 12: In situ hybridization of sections of _Styela plicata_ eggs
    with a [^{125}I]-labeled cDNA probe made from total egg polyA$^+$ RNA.
    A) A fertilized egg undergoing ooplasmic segregation showing
    grains concentrated over the ectoplasm (EC), which surfaces at
    several positions in the section. Some grains are present in the
    endoplasm (EN), but few grains are present in the yellow crescent
    region (YC). B) A two-cell embryo showing abundant grains in the
    ectoplasm but only a few grains in the yellow crescent cytoplasm
    (YC) of each blastomere. X 400.

general cytoplasm because of its high concentration of endoplasmic
reticulum and particulate materials.

     The distribution of individual mRNA sequences in the cytoplasmic
regions of _Styela_ embryos was also examined by _in situ_ hybridization
(Jeffery, 1982b; Jeffery et al., 1983). Information on the
distribution of histone and actin mRNA was obtained using
complementary DNA sequences from _Drosophila melanogaster_ as probes.
No differences were found in the concentration of grains developed
over the ectoplasm, endoplasm, or the yellow crescent cytoplasm after
in situ hybridization with the radioactively labeled histone DNA
probe (Fig. 13A-B). Quantification of these results showed that
about 10% of the total histone mRNA was present in the yellow
crescent cytoplasm. This value would be expected if histone messages
were homogeneously distributed between the yellow crescent (which
accounts for 11% of the egg volume; Jeffery and Capco, 1978), and the
other cytoplasmic regions. Histone mRNA is apparently representative
of a class of maternal messages that is evenly distributed in eggs.
The even distribution may be required to sustain histone synthesis in
each part of the embryo during the period of rapid cleavages.

     Unlike the situation observed for histone mRNA distribution,
high concentrations of grains were developed in the yellow crescent
and ectoplasm after _in situ_ hybridization with a radioactively
labeled actin DNA probe (Jeffery et al., 1983), (Fig. 13C-D).
Quantification of these results showed that the yellow crescent

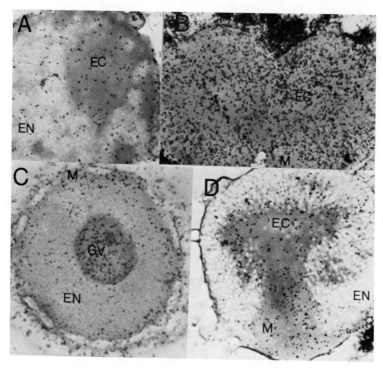

Figure 13: In situ hybridization of sections of Styela plicata eggs and embryos with [^{125}I]-labeled histone and actin DNA probes. The probes utilized for these experiments were a 4.6 kilobase Hind III fragment of the Drosophila melanogaster histone gene complex (Lifton et al., 1977) and a 1.8 kilobase Hind III fragment of the D. melanogaster DmA-2 actin gene (Fyrberg et al., 1980). A) An unfertilized egg probed with histone DNA. Grains are uniformly distributed between all three cytoplasmic regions. X 400. B) A two-cell embryo probed with histone DNA. Grains are uniformly distributed between all three cytoplasmic regions. The dark patches above the egg represent labeled follicle cells. X 400. C) A post-vitellogenic oocyte probed with actin DNA. Grains are concentrated in the germinal vesicle (GV) and the yellow cytoplasm (M) at the cell periphery. X 400. D) One of the blastomeres of a two-cell embryo probed with actin DNA. Grains are concentrated in the ectoplasm and the yellow crescent cytoplasm but not in the endoplasm. X 1,000. EC, ectoplasm; EN, endoplasm; M, yellow crescent cytoplasm. From Jeffery et al. (1983).

contains about 40% and the ectoplasm about 50% of the total actin mRNA, despite together accounting for only about 20% of the egg volume. The high concentration of actin mRNA in the ectoplasm is not surprising because almost half of the total mRNA is also concentrated in this region. The presence of a large proportion of the actin

messages in the yellow crescent, on the other hand, is unexpected
since this region is low in total mRNA. The results indicate that
much of the maternal actin mRNA is selectively localized in the
yellow crescent and segregated to the mesodermal lineage cells during
early embryogenesis.

What is the function of the localized maternal actin messages in
the yellow crescent? One possibility may be that the mesodermal
lineage cells require excess actin translational capacity during
early embryogenesis. Indeed, some of these cells begin to synthesize
and assemble myofibrils only 3-5 hr after fertilization. Jeffery
et al. (1983) have recently tested the possibility that mRNA coding
for the muscle-type actin species is stored in the yellow crescent.
The in vitro translation products of mRNA from Styela eggs and adult
siphon muscle tissue were compared by two-dimensional gel
electrophoresis. The muscle mRNA coded for three forms of actin,
including the most acidic form found only in muscle (Fig. 14). In
contrast, egg mRNA coded for only two forms of actin, both being the
more basic varieties found in all tissues. Even after long exposure
times, no muscle actin could be detected among the translation
products of the egg mRNA. The actin mRNA localized in the yellow
crescent apparently codes for one or both of the cytoplasmic forms of
actin.

Although the in situ hybridization approach has yielded valuable
information on the cytoplasmid distribution of selected mRNA species,
it cannot be conveniently used to assess the overall representation
of various mRNA species in the yellow crescent. Separating the
crescent from other parts of the egg would provide the most direct
solution to this problem. A procedure for the isolation of the
yellow crescent from Styela eggs has recently been developed by
Jeffery (unpublished). It is based on the previous finding that the
yellow crescent region is enriched in filamentous structures,
including F-actin filaments (Jeffery and Meier, 1983). Cytoskeletal
elements such as these are known to be very insoluble in high ionic
strength media (Cooke, 1976). Eggs were permeabilized by treatment
with low levels of Triton X-100 and exposed to buffered media
containing 0.5M KCl. After a few minutes, they were gently
homogenized and centrifuged. The pellet contained crude yellow
crescents, which appear to be resistant to homogenization. The
crescents can be isolated and purified by several cycles of
centrifugation through sucrose. An example of an isolated crescent
is shown in Fig. 15A. Electron microscopy showed that the isolated
crescents were quite similar in structure to the crescent region of
the intact egg (Fig. 15B). They contain pigment granules with
adherent mitochondria, endoplasmic reticulum, ribosome-like
particles, and the portion of plasma membrane which lies immediately
over the crescent in vivo.

Although the analysis of mRNA from the isolated yellow crescents

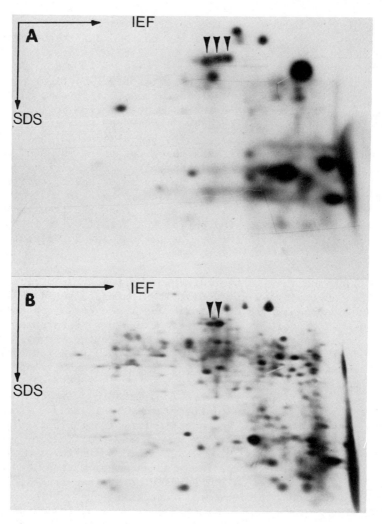

Figure 14:   Two-dimensional polyacrylamide gel electrophoresis of
^{35}S-methionine-labeled translation products synthesized in a wheat
germ lysate directed by RNA from siphon muscle (A) and eggs (B) of
Styela plicata. The arrows indicate the position of various actin
isoforms. The muscle actin isoform is the most acidic of the three
actin spots in A. From Jeffery et al. (1983).

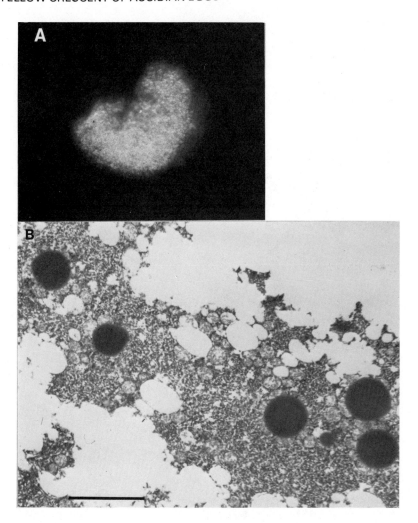

Figure 15: Light and electron micrographs of isolated yellow
  crescents from fertilized Styela plicata eggs. A) An isolated
  yellow crescent showing refractile pigment granules. X 320. B) An
  electron micrograph of an isolated yellow crescent showing pigment
  granules with adherent mitochondia. The bar represents 10
  microns.

is still in progress, interesting results have already been obtained.
RNA was extracted from the isolated crescent fraction, the
supernatant fraction, and intact eggs. After translation in a
cell-free system, the products were compared by two-dimensional gel
electrophoresis (Fig. 16). Most of the translation products of RNA
from the various fractions were qualitatively similar to those of

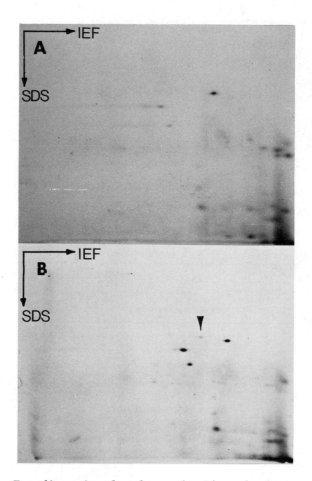

Figure 16: Two-dimensional polyacrylamide gel electrophoresis of
^{35}S-methionine labeled translation products synthesized in a wheat
germ lysate directed by RNA from the isolated yellow crescent (A)
or the supernatant (B) egg (ectoplasm and endoplasm) fraction of
Styela plicata eggs. A major egg polypeptide undetected in the
translation products of yellow crescent RNA is shown by the arrow.

intact eggs; however, some quantitative differences could be detected
between the gels. There was at least one translation product, an
80,000 molecular weight polypeptide, which was translated by total
egg and supernatant (ectoplasm plus endoplasm) mRNA, but was
undetectable in the translation products of the yellow crescent mRNA.
At this point in the study, however, it appears that most of the
prevalent messages of the yellow crescent are also present in other
parts of the egg. Similar results have been reported in two other
embryonic systems showing precocious localization of developmental

potential, the micromeres of sea urchins (Tufaro and Brandhorst, 1979), and the polar lobe of Ilyanassa (Brandhorst and Newrock, 1981; Collier and McCarthy, 1981).

In summary, current progress on the mRNA composition of the yellow crescent indicates the existence of a quantitative localization of sequences such as actin messages, but at present there is no evidence for strict qualitative localization, at least for the prevalent classes of messages. It is still possible, however, that yellow crescent region may contain a specific population of rare mRNA species or prevalent messages coding for basic proteins.

Proteins of the Yellow Crescent

Very little is currently known about the proteins of the yellow crescent. Berg and Baker (1962), described four antigens from Styela eggs. These antigens were found to be common to both the anterior and posterior blastomeres of four-cell embryos when they were examined by immunodiffusion methods. Recently, Jeffery (unpublished) has compared the major polypeptides of intact eggs and isolated yellow crescents by two-dimensional gel electrophoresis. The pattern of polypeptides in the yellow crescent appeared to be a distinct subset of those in the total egg (Fig. 17). Some of the major yellow crescent polypeptides can be positively identified. They include two iso-forms of actin, and the vimentin- and desmin-like intermediate filament proteins previously described by Jeffery and Meier (1983). Currently, it is unknown whether these and the other major yellow crescent proteins are also found in the remainder of the egg.

Interaction of Yellow Crescent mRNA with the Cytoskeleton

One of the most remarkable qualities of yellow crescent mRNA, which it shares with the other mRNA populations present in Styela eggs, is its resistance to mixing during ooplasmic segregation. In situ hybridization experiments have shown that the mRNA concentrations of the three major cytoplasmic regions remain unaltered after fertilization, during their extensive migration through the egg (Jeffery and Capco, 1978; Jeffery et al., 1983). Three reasons can be envisioned for the integrity of mRNA patterns during ooplasmic segregation. First, certain mRNA sequences could prefer local ionic microenvironments which are not modified during ooplasmic segregation. Second, the mRNA sequences could be associated with regionalized membranous systems or membrane-bound organelles in the various cytoplasmic regions. Third, the mRNA sequences could interact with localized cytoskeletal elements or organelles attached to the cytoskeleton, as seems to be the case in certain somatic cells (Lenk et al., 1977; Cervera et al., 1981; including ascidian follicle cells, Jeffery, 1982b).

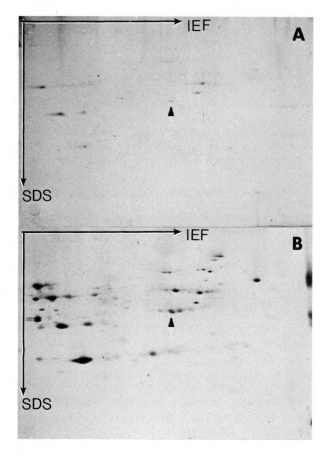

Figure 17:   Two-dimensional gel electrophoresis of protein present in
  the isolated yellow crescent fraction (A) and whole eggs (B) of
  Styela plicata. The position of actin is indicated by arrows.   The
  proteins are stained with silver.

        The association of mRNA with the egg cytoskeleton has recently
been tested by Jeffery (1983a). He reasoned that if mRNA is attached
to cytoskeletal elements, it should be resistant to extraction by
non-ionic detergents such as Triton X-100. Because these detergents
disrupt membranous systems, and would also be expected to destroy any
local ionic microenvironments, the retention of mRNA in detergent
extracted eggs should exclude these possibilities. When Styela eggs
are extracted with Triton X-100, about 60% of the egg poly(A) is left
in the detergent-insoluble fraction although only 20% of the total
proteins, 30% of the total RNA and less than 8% of the lipid (most of
this is due to the retention of lipid pigment granules bound to the
filamentous lattice in the residue) is present (Jeffery and Meier,
1983). These results suggest that egg mRNA is associated with

cytoskeletal elements. If this interpretation is correct, and the
association between mRNA and the detergent-insoluble material exists
in vivo, the concentration of mRNA in cytoskeletal residues of the
various cytoplasmic regions of the egg would be expected to be the
same as that of the intact cytoplasms. Jeffery (1983a) tested this
possibility by subjecting sections of Triton X-100-extracted eggs and
embryos to in situ hybridization with some of the probes discussed
earlier. The different regions of the detergent-extracted ascidian
egg can be distinguished by their staining properties in histological
preparations. Furthermore, the yellow crescent region of
detergent-extracted eggs can be precisely identified by the presence
of embedded pigment granules. In these experiments, the egg or
embryo population is divided into two parts; one part is extracted
with Triton X-100 while the other remains untreated. The
detergent-extracted and intact specimens are then mixed, fixed and
processed together for in situ hybridization. In this way the
distribution of grains in the different cytoplasms and their
cytoskeletal residues can be compared in a quantitative fashion.
When this experiment was conducted using the poly(U) and the actin
DNA probes, it was found that more than 75% of the total mRNA and
actin mRNA of the intact eggs was retained in various cytoplasmic
regions of their detergent-extracted counterparts. Moreover, both
total mRNA and actin mRNA molecules were concentrated in the expected
cytoplasmic regions of the detergent-extracted eggs (Fig. 18). The
total mRNA was present primarily in the ectoplasmic cytoskeletal
residue, while actin mRNA was concentrated in both the ectoplasmic
and yellow crescent cytoskeletal residues. The results suggest that
mRNA localization in the yellow crescent and other cytoplasmic
regions is based on the association of mRNA molecules with
regionalized cytoskeletal elements, their associated organelles, or
other detergent-insoluble structures. It will be an important
endeavor to determine the identity of these detergent-insoluble
structures and the recognition signals which must exist in maternal
mRNA.

Mechanism of Yellow Crescent Localization

     The mechanism of yellow crescent localization is of fundamental
importance to our understanding of how cytoplasmic determinants are
properly positioned in the egg. Accordingly this topic has been
actively investigated during the last decade. Two important
conclusions have been drawn from these studies. The first is that
structural components of the egg plasma membrane appear to
participate in ooplasmic segregation and "co-cap" with the underlying
yellow crescent pigmentation in the vegetal pole region of the egg.
An initial indication of this movement was obtained by Conklin
(1905a), who observed that the test cells, accessory cells closely
applied to the external surface of the egg, usually segregate in
unison with the yellow cytoplasm and come to rest over the yellow
crescent after the completion of ooplasmic segregation. The same is

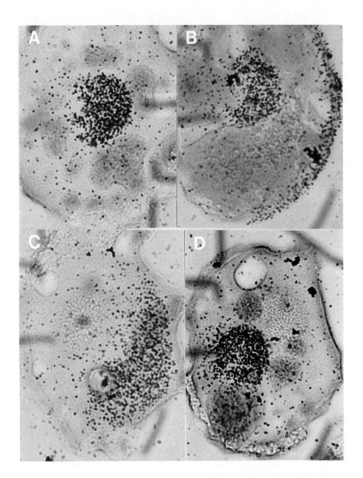

Figure 18:   In situ hybridization of sections of Triton X-100
extracted Styela oocytes and eggs with [^{3}H]-poly(U) and
[^{125}I]-actin DNA probes. A) A post-vitellogenic oocyte probed with
poly(U). Grains are concentrated in the area originally occupied
by the germinal vesicle. B) A post-vitellogenic oocyte probed with
actin DNA. The grains are concentrated in the areas originally
occupied by the GV (central), and the yellow cytoplasm (periphery).
Compare this autoradiograph with Fig. 13C. C) A fertilized egg
probed with poly(U). Grains are concentrated in the area
originally occupied by the ectoplasm. D) A fertilized egg probed
with actin DNA. The grains are concentrated in the areas
originally occupied by the ectoplasm (center), and the yellow
crescent (out of focus below the ectoplasm). X 400. From Jeffery
1983a).

true of supernumerary sperm which stick to the egg surface (Sawada and Osanai, 1981), carmine (Sawada and Osanai, 1981), and chalk (Ortolani, 1955b) particles applied to the egg, and fluorescent lectins which bind to surface macromolecules (Monroy et al., 1973; O'Delle et al., 1974; Ortolani et al., 1977; Zalokar, 1980). Substances bound to the cell surface are known to migrate to the yellow crescent region from as far away as the animal pole. This led Sawada and Osanai (1971) to propose that the egg cortex contracts uniformly toward the vegetal pole during ooplasmic segregation.

The second conclusion is that cytoskeletal elements are involved in ooplasmic segregation. Microtubules seem to be excluded because colchicine does not interfere with yellow crescent formation (Zalokar, 1974). Studies using cytochalasin B, however, have implicated actin microfilaments in ooplasmic segregation (Zalokar, 1974; Reverberi, 1975; Sawada and Osanai, 1981). As mentioned earlier, an electron dense layer (Sawada and Osanai, 1981), probably consisting of a network of actin filaments (the PML; Jeffery and Meier, 1983), is present beneath the plasma membrane of Styela and Boltenia eggs. This layer becomes thicker (Sawada and Osanai, 1981) and more densely packed with filaments (Jeffery and Meier, 1983) as yellow crescent segregation proceeds, suggesting that cytoskeletal contraction may be a motive force for ooplasmic segregation.

Based on these conclusions, Jeffery and Meier (1983) have recently proposed a model for the segregation of yellow crescent cytoplasm and its molecular constituents in ascidian eggs (Fig. 19). The model has several key elements. First, a cytoskeletal lattice-work in the yellow crescent region connects pigment granules, mitochondria, other organelles and certain molecular constituents (such as mRNA) to the actin network of the PML. The ultrastructural studies of Jeffery and Meier (1983) provided direct evidence for such a linkage, whereas in situ hybridization studies with detergent-extracted eggs (Jeffery, 1983) suggest that mRNA is associated with cytoskeletal elements or attached organelles in the yellow crescent region. Second, the PML is postulated to interact with integral membrane proteins, as reticula of this nature appear to do in other cells (Ben-Ze'ev et al., 1979; Sheetz, 1979). Finally, the PML is assumed to be a contractile structure.

According to the model, the initial event of ooplasmic segregation would be a uniform contraction of the PML toward a specific site in the vegetal hemisphere. The PML would continue to shorten and pull the yellow crescent cytoplasm (on its inner surface) and integral membrane proteins (on its outer surface), with it as ooplasmic segregation proceeds. The accumulation of excess membrane at the vegetal pole (see Fig. 4), may be the reason for the appearance of microvilli in this region. The contraction of the PML could also constitute the force that displaces the endoplasm into the animal hemisphere. The latter, coupled with a weakening of animal

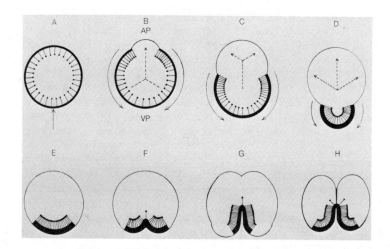

Figure 19: A model for the mechanism of ooplasmic segregation and
    yellow crescent localization in ascidian eggs. In each diagram the
    thick egg boundaries represent parts of the membrane containing the
    plasma membrane lamina (PML). The thin egg boundaries represent
    parts of the plasma membrane without the PML. The structures
    attached to the inside of the PML represent the internal
    filamentous lattice and associated components in the yellow
    crescent cytoskeleton, such as the pigment granules and
    mitochondria. A) An unfertilized egg. The arrow indicates the
    focal point on the egg surface for yellow crescent localization.
    B-D) Fertilized eggs involved in ooplasmic segregation. AP, animal
    pole; VP vegetal pole. The arrows with unbroken lines represent
    the direction of PML contraction. The arrows with broken lines
    represent the direction of endoplasmic movement. E) An egg that
    has completed ooplasmic segregation. F-H) Bisection of the yellow
    crescent at first cleavage. In part from Jeffery and Meier
    (1983).

pole plasma membrane (possibly by the depletion of integral
proteins), may lead to the bulging of the cytoplasm in the animal
hemisphere, whereas continued shortening of the PML and endoplasmic
bulging at the opposite end of the cell could ultimately cause a
cytoplasmic lobe to appear at the vegetal hemisphere of the egg (as
is often seen during ooplasmic segregation; Fig. 8; Reverberi, 1961;
Zalokar, 1974; Sawada and Osanai, 1981; Jeffery, 1982a). Further
studies on the organization of the yellow crescent cytoplasm,
particularly its cytoskeleton and plasma membrane, will be necessary
to determine whether the features of this model are correct.

An important problem which remains to be considered is how the migration of the yellow crescent cytoplasm is polarized during ooplasmic segregation. It has been previously suggested that ooplasmic segregaton and the distribution of cytoplasmic macromolecules in Fucus eggs are directed by a local increase in calcium ion concentration (reviewed by Jaffe, 1981). Evidence has recently been obtained that a calcium ion flux may also direct the migration of the yellow crescent cytoplasm in ascidian eggs (Jeffery, 1982a). The relation between the calcium flux and yellow crecent formation was tested by aligning Boltenia eggs against a small glass rod coated with the calcium ionophore A23187. A steep gradient of ionophore emanates from the glass rod (Robinson and Cone, 1980). The majority of the eggs aligned against the ionophore-coated rod form an orange crescent on the side of the cell facing the rod. Ionophore A23187 is known to activate ascidian eggs (Steinhardt et al., 1974; Bevan et. al., 1977) and induce orange crescent formation in eggs of Boltenia. Thus, the crescent seems to be polarized in the direction of highest calcium concentration. Moreover, if unfertilized eggs are simultaneously positioned between two ionophore coated glass rods, two crescents are formed, each about half the size of a normal crescent and localized precisely at the points where the rods touch the cell surface (Jeffery, 1982a). During normal development, the focus of yellow crescent polarization is likely to be determined by the point of sperm entry. As discussed earlier, Conklin (1905a) observed that the sperm enters the egg in the vegetal hemisphere. The problem of yellow crescent polarization in ascidian eggs is thus reduced to the question of how the fertilizing sperm recognizes the vegetal region of the egg.

CONCLUSIONS

As shown in this brief review, considerable information has been obtained on the ascidian yellow crescent since it was first described by Conklin in 1905. The structure of this cytoplasmic region has been worked out in some detail and excursions are also being made into its macromolecular constitution. The existence of a complex cytoskeletal organization in the yellow crescent and its function in mRNA localization provides the first link between gene expression and the cytoarchitecture of the egg. It has been demonstrated that the cytoskeletal organization, probably coupled to a calcium-based regulatory system, is also of crucial importance in directing the overall localization of the yellow crescent in early development. It is now almost certain that this region of the egg harbors cytoplasmic determinants that specify the development of the muscle and mesenchyme lineage cells. Unfortunately, the molecular nature of the yellow crescent cytoplasmic determinants is still pending; however, assay systems currently being developed provide some hope that this important problem will be solved in the near future.

ACKNOWLEDGEMENTS

The experiments described in this paper from our laboratories were supported by grants from the National Institutes of Health (GM-25119, HD-13970 and DE-05616), the National Sciences Foundation (PCM-791472), and the Muscular Dystrophy Association. Part of this work was also supported by National Institutes of Health training grant 323-HD-07098 for the Embryology course, Marine Biological Laboratory, Woods Hole. We are grateful to Ms. Bonnie Brodeur for assistance with the preparation of the figures and Ms. Ming Shi Chang, Priscilla Kemp, Dianne McCoig, Linda Wilson and Mr. Christopher Drake for technical assistance.

REFERENCES

Barak, L.S., Yocum, R.R., Nothnagel, E.A., and Webb, W.W. (1980). Fluorescence staining of the actin cytoskeleton in living cells with 7-nitrobenz-2-oxa-1, 3-diazole phallacidin. Proc. Natl. Acad. Sci. USA 77:980-984.

Ben-Ze'ev, A., Duerr, A., Solomon, F., and Penman, S. (1979). The outer boundary of the cytoskeleton: A lamina derived from plasma membrane proteins. Cell 17:859-865.

Berg, W.E. (1956). Cytochrome oxidase in anterior and posterior blastomeres of Ciona intestinalis. Biol. Bull. 110:1-7.

Berg, W.E. (1957). Chemical analysis of anterior and posterior blastomeres of Ciona intestinalis. Biol. Bull. 113:365-375.

Berg, W.E. and Humphreys, W.J. (1960). Electron microscopy of four-cell stages of the ascidians Ciona and Styela. Develop. Biol. 2:42-60.

Berg, W.E. and Baker, P.C. (1962). Antigens in isolated blastomeres of the ascidians Ciona and Styela. Acta Embryol. Morphol. Exp. 5:274-279.

Berrill, N.J. (1929). Studies in tunicate development. I. General physiology of development of simple ascidians. Phil. Trans. Roy. Soc. London, B, 218:37-78.

Berrill, N.J. (1932). The mosaic development of the ascidian egg. Biol. Bull. 63:381-386.

Berrill, N.J. (1968). Tunicata. In "Invertebrate Embryology," M. Kume and K. Dan (eds.), p. 538-576. National Library of Medicine, Washington D.C.

Bevan, S.J., O'Dell, D.S., and Ortolani, G. (1977). Experimental activation of ascidian eggs. Cell Differentiation 6:313-318.

Boyles, J. and Bainton, D.F. (1979). Changing patterns of plasma membrane-associated filaments during the initial phases of polymorphonuclear leukocyte adherence. J. Cell Biol. 82:347-368.

Brandhorst, B.P. and Newrock, K.W. (1981). Post-transcriptional regulation of protein synthesis in Ilyanassa embryos and isolated polar lobes. Develop. Biol. 83:250-254.

Brothers, A.J. (1979). A specific case of genetic control of early development: The o maternal effect mutation of the Mexican axolotl. In "Determinants of Spatial Organization," S. Subtelny and I.R. Konigsberg (eds.), p. 167-183. Academic Press, New York.

Cervera, M., Dreyfuss, G., and Penman, S. (1981). Messenger RNA is translated when associated with the cytoskeletal framework in normal and VSV-infected HeLa cells. Cell 23:1130-120.

Collier, J.R. and McCarthy, M.E. (1981). Regulation of polypeptide synthesis during early embryogenesis of Ilyanassa obsoleta. Differentiation 19:31-46.

Conklin, E.G. (1905a). The organization and cell lineage of the ascidian egg. J. Acad. Natl. Sci. Phil. 13:1-119.

Conklin, E.G. (1905b). Organ-forming substances in the eggs of ascidians. Biol. Bull. 8:205-230.

Conklin, E.G. (1905c). Mosaic development of ascidian eggs. J. Exp. Zool. 2:145-223.

Conklin, E.G. (1906). Does half an ascidian egg give rise to a whole larva? Roux Archiv. Entwicklungsmech. 21:727-753.

Conklin, E.G. (1931). The development of centrifuged eggs of ascidians. J. Exp. Zool. 60:1-119.

Cooke, P.A. (1976). Filamentous cytoskeleton in vertebrate smooth muscle cells. J. Cell Biol. 68:539-556.

Davidson, E.H. (1976). Gene Activity in Early Development, p. 245-318. Academic Press, New York.

DeSantis, R. and Stopak, D. (1980). Mitochondrial movements in early development of Ciona. Biol. Bull. 159:446.

Durante, M. (1956). Cholinesterase in development of Ciona
    intestinalis (Ascidia). Experientia 12:307-310.

Fyrberg, E.A., Kindle, D.L., Davidson, N., and Sodja, A. (1980).
    The actin genes of Drosophila: A dispersed multigene family.
    Cell 19:365-378.

Hainfied, J.F. and Steck, T.L. (1977). The sub-membrane reticulum
    of the human erythrocyte: A scanning electron microscope study.
    J. Supramol. Struct. 6:301-311.

Harvey, L.A. (1927). The history of cytoplasmic inclusions of the
    egg of Ciona intestinalis during oogenesis and fertilization.
    Proc. Roy. Soc. London, B, 101:137-161.

Hitchcock, S.E., Carlsson, L., and Lindberg, U. (1976).
    Depolymerization of F-actin by deoxyribonuclease I. Cell
    7:531-542.

Hsu, W.S. (1962). An electron microscopic study of the origin of
    yolk in the oocytes of the ascidian Boltenia villosa. La
    Cellule 62:145-165.

Hsu, W.S. (1963). The nuclear envelope in the developing oocytes of
    the tunicate Boltenia villosa. Z. Zellforsch. 58:17-26.

Illmensee, K. and Mahowald, A.P. (1974). Transplantation of
    posterior polar plasm in Drosophila. Induction of germ cells at
    the anterior pole of the egg. Proc. Natl. Acad. Sci. USA
    71:1016-1020.

Jaffe, L.F. (1981). Calcium explosions as triggers of development.
    Ann. N.Y. Acad. Sci. 399:86-101.

Jeffery, W.R. (1982a). Calcium ionophore polarizes ooplasmic
    segregation in ascidian eggs. Science 216:545-547.

Jeffery, W.R. (1982b). Messenger RNA in the cytoskeletal framework:
    Analysis by in situ hybridization. J. Cell Biol. 95:1-7.

Jeffery, W.R. (1983a). Maternal mRNA localization in ascidian eggs:
    Involvement of the cytoskeleton. Submitted.

Jeffery, W.R. (1981). Messenger RNA localization and cytoskeletal
    domains in ascidian embryos. In "Time, Space, and Pattern in
    Embryonic Development," W.R. Jeffery and R.A. Raff (eds.). A.R.
    Liss, New York, in press.

Jeffery, W.R. and Capco, D.G. (1978). Differential accumulation and
    localization of maternal poly(A)-containing RNA during early
    development of the ascidian, Styela. Develop. Biol.
    67:152-166.

Jeffery, W.R. and Meier, S. (1983). A yellow crescent cytoskeletal
    domain in ascidian eggs and its role in early development.
    Develop. Biol., in press.

Jeffery, W.R., Tomlinson, C.R., and Brodeur, R.D. (1983).
    Localization and segregation of cytoplasmic-type actin messenger
    RNA during early ascidian development, submitted.

Kalthoff, K. (1979). Analysis of a morphogenetic determinant in an
    insect embryo (Smittia spec., Chironomidae, Diptera). In
    "Determinants of Spatial Organization," S. Subtelny and I.R.
    Konigsberg (eds.), p. 97-126. Academic Press, New York.

Kessel, R.G. and Beams, H.W. (1965). An unusual configuration of
    the Golgi complex in pigment-producing "test" cells of the ovary
    of the tunicate, Styela. J. Cell Biol. 25:55-67.

Lenk, R., Ranson, L., Kaufman, Y., and Penman, S. (1977). A
    cytoskeletal structure with associated polyribosomes obtained
    from HeLa cells. Cell 10:67-78.

Lifton, R.P., Goldberg, M.L., Karp, R.W., and Hogness, D.S. (1977).
    The organization of the histone genes in Drosophila
    melanogaster: Functional and evolutionary implications. Cold
    Spr. Harb. Sym. Quant. Biol. 42:1047-1051.

Mahowald, A.P., Allis, C.D., Karrer, K.M., Underwood, E.M., and
    Waring, G.L. (1979). Germ plasm and pole cells of Drosophila.
    In "Determinants of Spatial Organization," S. Subtelny and I.R.
    Konigsberg (eds.), p. 127-146. Academic Press, New York.

Mancuso, V. (1963). Distribution of the components of normal
    unfertilized eggs of Ciona intestinalis. Acta Embryol. Morph.
    Exp. 5:71-81.

Monroy, A., Ortolani, G., O'Dell, D., and Millonig, G. (1973).
    Binding of concanavalin A to the surface of unfertilized and
    fertilized ascidian eggs. Nature 242:409-410.

Morgan, T.H. (1934). Embryology and Genetics. Columbia Univ.
    Press, New York.

O'Dell, D.S., Ortolani, G., and Monroy, A. (1974). Increased
    binding of concanavalin A during maturation of ascidian eggs.
    Exp. Cell Res. 83:408-411.

Ortolani, G. (1955a). The presumptive territory of the mesoderm in the ascidian germ. Experienta 11:445-446.

Ortolani, G. (1955b). I movementi corticali dell uovo di Ascidie alla fecondaziona. Riv. Biol. 47:169-181.

Ortolani, G., O'Dell, D.S., and Monroy, A. (1977). Localized binding of Dolichos lectin to the early Ascidia embryo. Exp. Cell Res. 106:402-404.

Pisano, A. (1949). Lo sviluppo dei primi due blastomeri separati dell' uovo di Ascidie. Pubbl. Staz. Zool. Napoli. 22:16-25.

Pucci-Minafra, I. and Ortolani, G. (1968). Differentiation and tissue interaction during muscle development of ascidian tadpoles. An electron microscope study. Develop. Biol. 17:692-712.

Raju, T.R., Stewart, M., and Buckley, I.K. (1978). Selective extraction of cytoplasmic actin-containing filaments with DNA-ase I. Cytobiol. 17:307-311.

Reverberi, G. (1956). The mitochondrial pattern in the development of the ascidian egg. Experientia 12:55-60.

Reverberi, G. (1961). The embryology of ascidians. Adv. Morph. 1:55-101.

Reverberi, G. (1975). On some effects of cytochalasin B on the eggs and tadpoles of ascidians. Acta Embryol. Exp. 2:137-158.

Reverberi, G. and Pitotti, M. (1939). Differenziazione fisiologiche nell' uovo della Ascidie. Pontif. Acad. Sci. Comment. 3:469-488.

Reverberi, G. and Minganti, A. (1946). Fenomeni di evocazione nello sviluppo dell' uovo di Ascidie. Risultati dell' indagine spermimen-tale sull' uovo di Ascidiella aspersa e di Ascidia malaca. Pubbl. Staz. Zool. Napoli. 16:363-401.

Ries, E. (1937). Die Verteilung von Vitamin C, Gluthation, Benzidin Peroxydase, Phenolase and Leukomethylenblau-Oxydoreductase wahrend der fruhen Embryonalentwicklung verschiedener wirbelloser Tiere. Pubbl. Staz. Zool. Napoli. 16:363-401.

Robinson, K.R. and Cone, R. (1980). Polarization of fucoid eggs by a calcium ionophore gradient. Science 207:77-78.

Sawada, T. and Osanai, K. (1981). The cortical contraction related to ooplasmic segregation in Ciona intestinalis eggs. Wilhelm Roux' Archiv. 190:208-214.

Sheetz, M. (1979). Integral membrane protein interaction with Triton cytoskeletons of erythrocytes. Biochem. Biophys. Acta 557:122-134.

Spek, J. (1926). Uber gesetzmassige Substanverteilungen bei der Durchung des Ctenophoreneies und ihre Beziehungen zu den Determinations Problemen. Roux. Arch. Entwicklungsmech. 107:54-73.

Steinhardt, R.A., Epel, D., Carroll, E.J., and Yanagamachi, R. (1974). Is calcium ionophore a universal activator for unfertilized eggs? Nature 252:41-43.

Tufaro, F. and Brandhorst, B.P. (1979). Similarity of proteins synthesized by isolated blastomeres of sea urchin embryos. Develop. Biol. 72:390-397.

Tung T. (1934). Recherches sur les potentialites des blastomeres chez Ascidella scabra. Experiences de translocation, de combinaison et d'isolement de blastomeres. Arch. Anat. Micros. 30:381-410.

Tung, T., Ku, S., and Tung, Y. (1941). The development of the ascidian egg centrifuged before fertilization. Biol. Bull. 80:153-168.

von Ubisch, L. (1940). Weitere Unterschungen uber Regulation und Determination im Ascidienkeim. Roux. Arch. Entwicklungsmech. 140:1-24.

Ursprung, H. and Schabtach, E. (1964). The fine structure of the egg of Ascidia nigra. J. Exp. Zool. 156:253-268.

Whittaker, J.R. (1973). Segregation during ascidian embryogenesis of egg cytoplasmic information for tissue-specific enzyme development. Proc. Natl. Acad. Sci. USA 70:2096-2100.

Whittaker, J.R. (1977). Segregation during cleavage of a factor determining endodermal alkaline phosphatase development in ascidian embryos. J. Exp. Zool. 202:139-154.

Whittaker, J.R. (1979). Development of tail muscle acetylcholinesterase in ascidian embryos lacking mitochondrial localization and segregation. Biol. Bull. 157:344-355.

Whittaker, J.R. (1980). Acetylcholinesterase development in extra
    cells caused by changing the distribution of myoplasm in
    ascidian embryos. J. Embryol. Exp. Morphol. 55:343-354.

Whittaker, J.R. (1982). Muscle lineage cytoplasm can change the
    developmental expression in epidermal lineage cells of ascidian
    embryos. Develop. Biol. 93:463-470.

Wilson, E.B. (1925). The Cell in Development and Heredity, 3rd
    edition, p. 1035-1121. MacMillan, New York.

Zalokar, M. (1974). Effect of colchicine and cytochalasin B on
    ooplasmic segregation of ascidian eggs. Wilhelm Roux' Archiv.
    175:243-248.

Zalokar, M. (1980). Effect of cell surface binding on development
    of ascidian egg. Wilhelm Roux' Archiv. 187:35-47.

EXPRESSION OF MATERNAL AND EMBRYONIC

GENES DURING SEA URCHIN DEVELOPMENT

Bruce P. Brandhorst, Frank Tufaro, and
Pierre-Andre Bedard[*]

ABSTRACT

We have investigated the patterns of protein synthesis during
embryonic development using two-dimensional electrophoresis. The
recruitment of a large amount of new maternal mRNA into polysomes
following fertilization is accompanied by very few changes in the
relative rates of synthesis of about 1000 polypeptides detected.
Between the time of hatching and invagination, many polypeptides
become undetectible and many others appear. A variety of other
developmental changes occur in this period as well, indicating that a
key transition has occurred from early development, dominated by
expression of the maternal genome to late development, which requires
participation of the embryonic genome. About 20% of the newly
synthesized polypeptides, mostly minor ones, undergo changes in
relative rates of synthesis during development, and far fewer change
detectibly in mass. A variety of patterns of metabolism are observed
for different polypeptides. Cloned cDNAs have been identified which
correspond to mRNAs coding for some of these developmentally
regulated proteins. Some of the newly synthesized polypeptides are
enriched in one of the three primary tissue layers, and these
increase in relative rates of synthesis during development. A
comparison of polypeptides enriched in ectoderm or endoderm
demonstrates little evolutionary conservation among three genera.
Few prevalent paternal mRNAs are detectible in comparable abundance
in interspecies hybrid embryos even at the mature pluteus stage. For
some genes this may be because the stored maternal RNA persists

---

[*]From the Department of Biology, McGill University, Montreal,
Quebec H3A 1B1, Canada.

throughout embryonic development without being replenished by new synthesis. We have identified several cloned DNAs complementary to transcripts which normally accumulate extensively during embryogenesis of the paternal species but are barely detectible in the hybrid embryos. The possible defects in expression of these paternal genes in foreign cytoplasm are discussed.

INTRODUCTION

The development of a complex organism from a fertilized egg most certainly requires elaborate temporal and spatial manifestations of differential gene activity. Nevertheless, the extent to which this is so has remained vague in spite of recent advances in methodology which provide elegantly detailed information about the structure and function of particular genes. In this report we summarize a variety of recent investigations of sea urchin embryos and make some generalizations about the extent to which gene expression changes during embryonic development, the timing and tissue specificity of these changes, and the possible levels of regulation of these changes. We also consider the respective roles of the maternal and embryonic genomes which function during embryonic development. These investigations have allowed us to identify and, in some cases, clone the DNA of genes exhibiting interesting activity during embryogenesis.

Following fertilization, the sea urchin egg undergoes a period of very rapid cleavage. The fourth cleavage gives rise to blastomeres of three different sizes and developmental fates: macromeres, mesomeres, and micromeres. Micromeres cultured in isolation differentiate in situ into primary mesenchyme cells which ultimately secrete skeletal calcitic spicules (Okazaki, 1975); they are thus determined and require no further interactions with other blastomeres to express this state of determination. Subsequent cleavages give rise to a blastula which ultimately hatches from the fertilization envelope. Primary mesenchyme cells begin to appear in the blastocoel a few hours after hatching, and gastrulation by invagination begins a few hours later. Following completion of gastrulation, skeletal formation is rapid, radial symmetry is lost, and the bilaterally symmetric pluteus larva forms. The pluteus is highly complex, consisting of a variety of differentiated tissues. Embryonic development to the feeding pluteus stage takes a few days, depending on the temperature and species.

Activation of Translation of Stored Maternal mRNA

The sea urchin egg contains proteins, RNA, and other materials required for embryonic development, there being no net change in mass until the pluteus begins to feed. In particular, sea urchin eggs contain a store of translationally inactive maternal mRNA, much of which quickly enters polysomes following fertilization (Gross et al., 1964; Humphreys, 1969, 1971). While this maternal mRNA accounts for most of the protein synthesis during early development, mRNA is also actively synthesized within a few hours after fertilization; ultimately, mRNA transcribed from the embryonic genome replaces the maternal mRNA. The timing of these events is considered later in this review. We have been interested in defining the respective roles of these two classes of mRNA functioning in the embryo.

Maternal mRNA might be expected to code for a special set of proteins required for cleavage or other events in early embryonic development. To test this hypothesis we compared the proteins synthesized by the unfertilized egg with those of zygotes or early embryos. Cultures were incubated with ^{35}S-methionine, lysates were prepared, and proteins were resolved by two-dimensional electrophoresis according to O'Farrell (1975). As shown in Fig. 1, comparison of approximately 10,000 individual polypeptides showed very few differences in extent of labeling before and after fertilization in spite of the greater than 20-fold increase in amount of mRNA being translated (Brandhorst, 1976; Tufaro and Brandhorst, 1979). As applied, this technique detects only those polypeptides having isoelectric points between about 4.8 and 7.1 and molecular weights between about 10 and 125 k daltons; most of the mass of newly synthesized protein of eggs and zygotes falls within this range (Brandhorst, 1976), though some proteins, such as histones, are notably excluded. The sequence complexity of RNA in polysomes of embryos suggests that we detect only about one-tenth of the translation products expected (Galau et al., 1974; Hough-Evans et al., 1977; Wilt, 1977; Duncan and Humphreys, 1981a; Tufaro and Brandhorst, 1982). Presumably, we detect only the products of the more abundant and/or efficiently translated mRNAs (this is discussed further below). In spite of these limitations it is clear that the activation of stored maternal mRNA involves, for the most part, a quantitative rather than qualitative change in the population of mRNA being translated; specific mRNA sequences are not being selected for translation.

Most, and probably all, of the small amount of RNA being translated in the unfertilized egg is newly synthesized and unstable (Brandhorst, 1980). The RNA which is recruited into polysomes following fertilization was synthesized during oogenesis and stored in a stable, untranslatable state in the egg. Thus the same mRNA sequence can exist simultaneously in the same cell in two different states: translatable and unstable or untranslatable and stable. While these two classes of RNA are synthesized at different times, it is presently not known how they are distinguished. The most favored hypothesis is that proteins associated with the stored maternal mRNA prevent its translation in the egg (see Raff and Showman, 1983, for review).

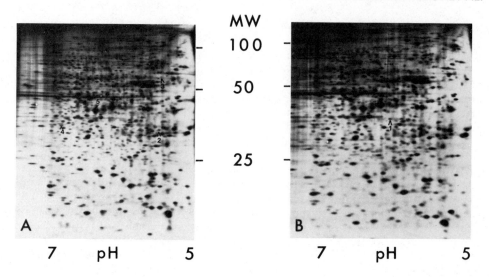

Figure 1: Comparison of proteins synthesized by unfertilized eggs
and 16-cell embryos. Eggs and embryos of S. purpuratus were
labeled with ³⁵S-methionine; proteins were extracted and separated
by two-dimensional electrophoresis. The autoradiographs show 997
distinct spots. Spot 1 was observed only in the embryos while
spots 2-4 were detectable only in the eggs. Spot 5 was observed in
these eggs and embryos but not observed in several other batches of
eggs and embryos. (A) eggs; (B) embryos. Reproduced from Tufaro
and Brandhorst (1979), with permission of Academic Press, Inc.

     Not all maternal mRNA sequences are recruited into polysomes
shortly after fertilization. Some stored histone mRNAs do not begin
to enter polysomes until just before the first cleavage (Wells et
al., 1981). These appear to be sequestered in or near the pronucleus
(Showman et al., 1982; Venezky et al., 1981). One polypeptide not
detected in eggs, but coded by maternal RNA, accumulates only during
part of the cell cycle and is probably rapidly degraded at other
times (Evans et al., 1982).

     When oocytes of another echinoderm, the starfish, are induced by
1-methyladenine to undergo meiotic maturation, many changes in
pattern of protein synthesis occur; this requires changes in the
population of maternal mRNA available for translation (Rosenthal et
al., 1982). Specific maternal mRNAs present but not translated in
oocytes appear in polysomes about the time of germinal vesicle
breakdown. These stored maternal mRNAs are not sequestered in the
germinal vesicle, nor is the release of its contents required for
their translational activation (Martindale and Brandhorst, 1982). As
in sea urchins, fertilization of the starfish egg does not result in

a further change in the pattern of protein synthesis. Changes in patterns of protein synthesis are commonly associated with meiotic maturation of oocytes in a wide variety of organisms (Rosenthal et al., 1982). We predict that meiotic maturation of sea urchin eggs is also associated with a change from a pattern of protein synthesis characteristic of oocytes to the pattern characteristic of eggs and early embryos.

Patterns of Protein Synthesis During Embryonic Development

Substantial changes in the patterns of protein synthesis occur during later embryonic development in sea urchins. For example, Fig. 2 shows a comparison of proteins synthesized in eggs and plutei of Lytechinus pictus. Polypeptide spots which have undergone changes in relative rates of synthesis are identified. Triangles pointing up indicate spots having increased intensity compared to the other stage; triangles pointing down indicate spots having decreased intensity compared to the other stage. Arrows point to spots detectible at that stage but absent at the other stage, for which the corresponding positions are shown by open circles. The definition of these "qualitative" changes is arbitrary. To normalize for the increasing absolute rate of protein synthesis (Goustin and Wilt, 1981), autoradiographic exposure was for the same total disintegrations of radioactive protein loaded onto each gel. Since longer exposure times increase the number of spots which are detectible in some areas of the gel (but decrease resolution in other areas), some apparently qualitative changes must be the result of limited sensitivity. The patterns of proteins synthesized by S. purpuratus embryos have been subjected to quantitative analyses (Bedard and Brandhorst, 1983). Only about 20% of the polypeptides undergo any detectible changes in relative rates of synthesis during embryonic development. Of these, about half change by at least 10 fold, but only about 1% of the spots change by 100 fold or more; for spots undergoing "qualitative" changes (about half of the total), these values are minimal estimates of the extent of change. No spot ever accounts for more than a few percent of the total protein synthesis of the embryo.

The timing of these changes has been analyzed by labeling embryos with ^{35}S-methionine for 1-2 hr at intervals throughout embryogenesis. In Fig. 3 the frequency with which spots have appeared or disappeared since the preceding labeling period ("qualitative" changes) have been plotted for each interval. Few changes in pattern of protein synthesis occur in the first few hours of development, but changes become much more frequent about the time of hatching. In fact, nearly all the polypeptides synthesized in eggs and early embryos which become undetectible during embryogenesis disappear between hatching and the completion of gastrulation. Many of the spots which become undetectible are barely detectible at any stage and their disappearance is sometimes transient; these

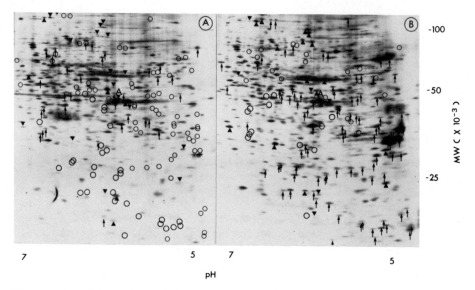

Figure 2: Comparison of proteins synthesized by eggs and plutei.
Eggs of L. pictus were incubated with ^{35}S-methionine for 4 hr;
plutei, 66 hr after fertilization, were incubated with
^{35}S-methionine for 1 hr. Proteins were extracted and separated by
two-dimensional electrophoresis as described by Bedard and
Brandhorst (1983); autoradiography was for 1.8 x 10^{10}
disintegrations applied to the first dimension. Spots showing
differences in relative intensities are marked as described in the
text. The spot labeled A is the major actin variant. The cluster
of spots labeled T are the tubulins.

apparently qualitative changes are thus probably the result of minor
variations in relative rates of synthesis or autoradiographic
sensitivity.

Appearances of new spots are most frequent in the interval
between hatching and gastrulation as well. The most striking
quantitative increases, both in extent and number, also occur during
the period between a few hours before hatching and the beginning of
gastrulation. Many of the spots which appear or increase in
intensity for the first time in this period continue to increase for
many hours thereafter. By pluteus stage nearly all intensely labeled
spots have undergone large increases in relative rates of synthesis.
Of the polypeptides showing changes of more than 100 fold, all but
one increase during development. Overall there is a modest increase
in the number of detectible, newly synthesized polypeptides. Most of
the polypeptides whose relative rates of synthesis change are never
among the prominent translation products. Thus the overall pattern

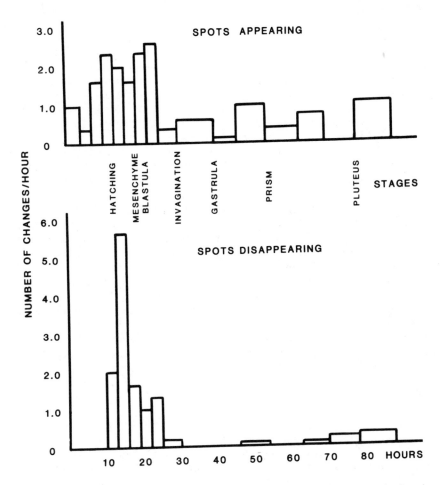

Figure 3: Timing of "qualitative" changes in protein synthesis during development of S. purpuratus embryos. Embryos were labeled with ^{35}S-methionine for 1-2 hr at various times throughout development. Proteins were extracted, separated by two-dimensional electrophoresis, and analyzed by autoradiography at exposures of 1.8×10^{10} disintegrations. The number of "qualitative" changes observed for each labeling period compared to the previous labeling period is plotted against time after fertilization and developmental stage. The upper panel refers to new spots detectible for the first time; the lower panel refers to spots which have become undetectible in that interval. The area included in each bar is equal to the number of spots which have changed in that interval. Note that fewer spots (about 800) were analyzed for this investigation than for that shown in Fig. 1 due to differences in sample preparation and autoradiographic exposure.

of protein synthesis is remarkably constant during embryonic
development.

These investigations of protein synthesis indicate that at about
the time of hatching, extensive, rapid changes in gene expression
begin. Hatching marks the beginning of a very interesting period in
development in other ways as well. Embryos cultured in the presence
of the inhibitor of RNA synthesis, Actinomycin D, arrest at the time
of hatching (Gross et al., 1964). An abrupt change in synthesis of
histone variants occurs around the time of hatching; it is mediated
by the replacement of early histone mRNAs by recently synthesized
late histone mRNAs (Newrock et al., 1978; Grunstein, 1978; Hieter et
al., 1979; Childs et al., 1979). The rate of cell division slows
considerably at the time of hatching such that there is no longer an
exponential increase in cell number (Hinegardner, 1967). By
mesenchyme blastula stage, a few hours after hatching, a large
fraction of the RNA being translated is transcribed from the
embryonic genome (Brandhorst and Humphreys, 1972; Cabrera et al.,
1983). Cultured primary mesenchyme cells begin to undergo extensive,
rapid changes in patterns of proteins synthesis at a time
corresponding to hatching (Harkey and Whiteley, 1982b). If N-linked
glycoprotein synthesis is inhibited by tunicamycin, gastrulation does
not occur (Schneider et al., 1978; Heifetz and Lennarz, 1979).
Synthesis of the intermediates in N-linked protein glycosylation,
dolichol and dolichyl phosphate, reaches a maximal rate a few hours
after hatching (Carson et al., 1982). Changes in patterns of protein
synthesis as a result of vegetalization induced by LiCl first appear
after hatching (Hutchins and Brandhorst, 1979).

Some or all of these developmental changes which are observed
between hatching and gastrulation might be triggered by a common
event or set of related events. Taken together, these observations
suggest that a rather abrupt developmental transition from early to
late embryonic development begins at the time of hatching. It is
possible that many of the changes in protein synthesis observed at
this time are coordinately regulated. A variety of changes in
cellular behavior and transcription also abruptly occur in Xenopus
embryos after the twelfth cleavage (Newport and Kirschner, 1982a);
the timing of these events appears to be related to achievement of a
critical ratio of nucleus to cytoplasm (Newport and Kirschner,
1982b).

We have investigated the possibility that changes in patterns of
protein synthesis observed during embryonic development might be
mediated by regulation of the availability for translation of
specific stored mRNA sequences. We compared the products of
cell-free translation of RNA extracted from embryos of various stages
to proteins synthesized in vivo using two-dimensional electrophoresis
(Bedard and Brandhorst, in preparation). In general, an mRNA
sequence translatable in wheat germ or reticulocyte lysates is

detectible only at stages when it is being translated in vivo; this observation applies to nearly all of those cell-free translation products (about half) which co-migrate with in vivo products. It remains possible that some mRNAs are stored in a form untranslatable in cell-free systems, as well as in vivo, but are ultimately processed to a translatable form (see below for more discussion of this possibility).

The arrest of embryos cultured in Actinomycin D at the hatching blastula stage suggests that active transcription is required for the developmental changes which begin at hatching. Until then, translation of maternal mRNA is apparently sufficient for normal, though retarded, development (Gross et al., 1964). The simplest interpretation of the observations presently available is that the rapid changes in patterns of protein synthesis occurring around the time of hatching are due to replacement of maternal mRNAs with the newly transcribed mRNA representing a distinct but overlapping set of genes. At the time of hatching there are several hundred blastomeres, so it is likely that newly synthesized mRNA accumulates rapidly enough to account for the observed increases in synthesis of polypeptides. The rapid decline in synthesis of many polypeptides following hatching may be due to exhaustion of the maternal mRNAs coding for them through normal decay.

Patterns of Protein Metabolism During Embryonic Development

Changes in mass of individual polypeptides were analyzed by comparing silver-stained two-dimensional electrophoretic separations of proteins of eggs and plutei (Bedard and Brandhorst, 1983). These patterns are even more similar than the patterns of newly synthesized proteins, but some important changes are obvious. Several very prevalent egg proteins decline greatly or are undetectible by pluteus stage. These proteins decline gradually in mass during embryonic development and most are not detectibly synthesized during embryogenesis. They are synthesized and stored during oogenesis and processed or degraded during embryonic development and presumably include the yolk proteins.

A few proteins accumulate substantially during development. These include a group of small acidic proteins, enriched in the ectoderm (Bruskin et al., 1982), whose relative rate of synthesis increases by about 200 fold. A variety of patterns of protein metabolism can be discerned by comparing stained gels with autoradiographs of gels loaded with newly synthesized radioactive proteins. Some spots are never detectibly labeled during embryonic development, but no decline in their mass is detected; while some of these may lack methionine, most are probably quite stable. These include many proteins which associate with chromatin during embryogenesis and are stored in the egg (Kuhn and Wilt, 1981). Many proteins are synthesized throughout embryonic development at a

constant relative rate of synthesis but do not change in mass;
presumably, their mass is maintained by turnover. The relative rates
of synthesis of other polypeptides increase or decrease considerably
during development, but they remain almost constant in mass. For
example, the major actin variant (β-actin; Durica and Crain, 1982)
increases in labeling intensity by about 100 fold during development
but does not change in mass. These observations indicate that
different polypeptides have quite different stabilities and that
these stabilities can change during development. Post-translational
events are probably critical components of the processes regulating
gene activity during development. The most striking feature of these
comparisons is the great similarity of proteins present in eggs and
plutei. This observation demonstrates the extent to which oogenesis
prepares the egg for embryonic development. This preprogramming
presumably minimizes the extent of differential gene regulation
required during development and allows for an accelerated rate of
embryogenesis.

Tissue Specificity of Protein Synthesis in Sea Urchin Embryos

At the 16-cell stage the micromeres are determined to form
primary mesenchyme even in isolation (Okazaki, 1975), but the
patterns of proteins synthesized by micromeres are indistinguishable
from those of mesomeres and macromeres (Tufaro and Brandhorst, 1979;
Harkey and Whiteley, 1982a). On the other hand, mesomeres/macromeres
have a distinctive population of complex class (rare) maternal mRNA
sequences compared to micromeres, but these are not represented on
polysomes (Rodgers and Gross, 1978; Ernst et al., 1980). Primary
mesenchyme cells isolated at the beginning of gastrulation, which are
derived from micromeres, have a pattern of protein synthesis quite
distinct from epithelial cells (presumptive endoderm and ectoderm;
Harkey and Whiteley, 1982b). Isolated micromeres in culture undergo
abrupt, extensive changes in protein synthesis at a time
corresponding to the period between hatching and invagination (Harkey
and Whiteley, 1982a); this is the period in the intact embryo when
the primary mesenchyme cells appear in the blastocoel by ingression
from the vegetal plate. Expression of the differentiated state of
these cells, the formation of skeletal spicules, does not begin until
several hours after these changes in protein synthesis, toward the
end of gastrulation. There are very few changes in their patterns of
protein synthesis during the period when cultured primary mesenchyme
cells secrete and elaborate skeletal elements (Harkey and Whiteley,
1982a). The timing of changes in patterns of proteins synthesis
during embryonic development suggest that critical changes in gene
expression occur substantially earlier than overt expression of
specialized morphology and function. On the other hand, the
determined state of micromeres is not reflected in a detectibly
unique pattern of protein synthesis. The silver-stained pattern of
proteins extracted from micromeres is identical to that of
mesomeres/macromeres as well (our unpublished observations).

Primary mesenchyme cells constitute only a small fraction of the mass of cytoplasm in the embryo, and they may not be very accessible to radioactive amino acid precursors of protein synthesis (Harkey and Whiteley, 1982b). Consequently, the synthesis of many of the mesenchyme-enriched proteins was probably not detectible in our investigations of protein synthesis in intact embryos. Thus, our observations on the timing of changes in protein synthesis probably underestimate the extent of change, particularly in the period following hatching. Unfortunately, the pH gradients of gels used in our investigations and those of Harkey and Whiteley (1982a, b) are quite different, making direct comparisons difficult and tentative.

By the completion of gastrulation, ectoderm and endoderm/mesoderm fractions can be conveniently separated (McClay and Chambers, 1978). As discussed above, most of the labeled proteins in the endoderm/mesoderm fraction are probably synthesized by the endoderm. Comparisons of protein synthesis in these two fractions by two-dimensional electrophoresis indicate that the vast majority of polypeptides are shared in S. purpuratus (Bruskin et al., 1982). As shown in Fig. 4, a few newly synthesized polypeptides are greatly enriched in the ectoderm or endoderm/mesoderm fractions of L. pictus. These proteins were labeled in intact embryos before separation of the tissues. The separation techniques employed can induce a stress response which leads to alterations in protein synthesis; some of the induced polypeptides are of the same sizes as those induced in heat shocked embryos (our unpublished observations).

The relative rates of synthesis of all of these tissue-enriched polypeptides increases substantially during development, though some are detectibly synthesized in eggs. Some of these tissue-specific proteins are well represented by mass in the egg; it is not yet known whether they are nonrandomly distributed in the egg cytoplasm or segregated during early cleavage. The relative rates of synthesis of some ectoderm specific polypeptides begin to increase during cleavage while the synthesis of endoderm specific proteins generally increases after the beginning of invagination (Bruskin et al., 1982; Bedard and Brandhorst, 1983; our unpublished observations). Cloned cDNAs corresponding to a family of ectoderm specific proteins have been identified and characterized (Bruskin et al., 1981, 1982; see also chapter by Klein et al., in this book). Recently, cloned cDNAs for tissue-enriched mRNAs have been identified for Tripneustes gratilla (Fregien et al., 1982) and L. pictus (Conlon and Brandhorst, unpublished observations).

With the exception of the tubulins which are enriched in ectodermal cilia, the function of these tissue-specific proteins is unknown. We have compared the newly synthesized polypeptides greatly enriched in ectoderm or endoderm/mesoderm of three widely divergent species: S. purpuratus, L. pictus, and Arbacia punctulata. Except for the tubulins, the tissue-specific polypeptides are not shared by

MW(X10⁻³)

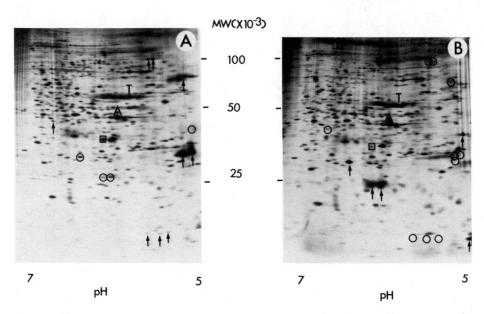

Figure 4: Comparison of proteins synthesized in the ectoderm (A) or
endoderm/mesoderm (B) of L. pictus early plutei. Embryos were
labeled with ^{35}S-methionine for 2 hr; ectoderm or endoderm/mesoderm
fractions were prepared according to McClay and Chambers (1978) and
Bruskin et al. (1982). Proteins were extracted and separated by
two-dimensional electrophoresis. Spots, or clusters of spots,
greatly enriched in that tissue are noted by arrows; circles
surround the corresponding area in the other gel. Spot A is the
major actin variant; T identifies the cluster of tubulins which are
enriched in the ectoderm. The ectoderm-enriched spot surrounded by
a square is sometimes not detectibly synthesized.

all three species, though there are some clusters of similar spots.
As shown in Table 1, about half of the newly synthesized polypeptides
detected on two-dimensional gels are shared between S. purpuratus and
L. pictus; thus the poor correspondence between tissue-enriched
polypeptides is very surprising and suggests that there are weak
constraints on changes in their structure during evolution.

    We are currently attempting to prepare monospecific antibodies
to these tissue-enriched polypeptides in order to investigate their
localization within the egg, embryo, and blastomeres. It is likely
that many of the other changes in protein synthesis observed during
development are related to cellular differentiation, but these will
probably only be localized by more sensitive techniques such as in
situ hybridization.

Table 1.   Detection of Synthesis of Paternal Proteins in Hybrid
Embryos

Hybrid Cross[a]		# of Common Spots[b]	Number of Paternal Spots					
			Cleavage[c]		Hatching[c]		Pluteus[c]	
Egg	Sperm		Obs.[d]	Exp.[e]	Obs.[d]	Exp.[e]	Obs.[d]	Exp.[e]
Sd	Sp	500	0	60	0	60	3	70
Sp	Sd	500	0	80	5	90	5	90
Sp	Lp	275	0	120	1	120	2	130

[a]Species used were S. purpuratus (Sp), S. droebachiensis (Sd), and L.
pictus (Lp).
[b]This represents the number of newly synthesized polypeptides common
to both species detected on two-dimensional gels. It was
determined by overlapping autoradiographs of labeled proteins of
the two species compared and by confirming the identity of these
spots on a gel containing a mixture of the proteins of both
species. Numbers are approximate and are limited by the
resolution of the gel having the poorest resolution. Only
proteins which could be consistently and unambiguously analyzed
were counted.
[c]Stage of embryonic development analyzed.
[d]The number of spots distinct to the paternal species observed among
proteins synthesized by hybrid embryos of that stage.
[e]The number of paternally distinct spots observed for embryos of the
paternal species of that stage and expected if all paternal genes
are expressed in hybrid embryos. Modified from Tufaro and
Brandhorst (1982); reproduced with permission of Academic Press,
Inc.

Changes in RNA Populations During Embryonic Development

Comparisons of rare polysomal mRNA prepared from sea urchin
embryos of various stages demonstrate that most mRNA sequences are
shared but that some are stage specific (Galau et al., 1976).
Virtually all mRNA sequences associated with polysomes of embryos are
represented in the stored maternal RNA, and there is a decline in the
complexity of polysomal and total RNA during development (Galau et
al., 1976; Hough-Evans et al., 1977). Surveys of recombinant cDNA
libraries indicate that most prevalent and moderately prevalent
transcripts are present in similar abundance in eggs and plutei,
though there appears to be a decline in abundance at intervening

stages (Lasky et al., 1980; Flytzanis et al., 1982). Cloned cDNAs complementary to transcripts which accumulate extensively during development have been identified (Bruskin et al., 1981, 1982; Crain et al., 1981; Merlino et al., 1981; Flytzanis et al., 1982; Fregien et al., 1982; our unpublished observations on an L. pictus cDNA library). Our observations on the extent of change in patterns of protein synthesis during development suggested that a greater fraction of the clones complementary to mRNA in cDNA libraries should correspond to transcripts which change in prevalence. While there is a variety of possible explanations for this apparent discrepancy, we favor the following. Most of the polypeptides undergoing changes in relative rates of synthesis during development are minor, never being intensely labeled. The most highly labeled proteins at gastrula stage of S. purpuratus are the major actin variant and two members of the family of small, acidic, ectodermal proteins (Bedard and Brandhorst, 1983). The moderately prevalent transcripts for these account for only about 0.05-0.1% of the mass of polyadenylated RNA in the cytoplasm, corresponding to about 100-200 copies per cell (Xin et al., 1982; Brandhorst, unpublished observations). The relative rates of synthesis of most polypeptides, including those undergoing changes in synthesis, are at least 10-100 times lower (Bedard and Brandhorst, 1983). If the mRNAs corresponding to these polypeptides are translated with efficiencies similar to those for actin and the ectoderm proteins, then they must be present in 1-20 copies per cell or fewer. Such rare transcripts are not detectable by the method usually employed to screen recombinant DNA libraries (Lasky et al., 1980). On the other hand, if they are represented in only a few blastomeres, they could be relatively prevalent in these cells and account for the local accumulation of large amounts of specialized proteins. Since there are many other rare polysomal RNA sequences having abundances similar to these estimates (Galau et al., 1974; Duncan and Humphreys, 1981a), one might expect a much larger number of spots detectable by two-dimensional electrophoresis. Increasing the pH or size range of the gels does not greatly increase the number of spots (our unpublished observations). We tentatively conclude that the mRNA sequences corresponding to the missing spots are translated less efficiently or might even be nonfunctional.

The relative constancy of the RNA population during embryonic development and formation of distinct tissue layers suggests that many of these sequences may be the transcription products of genes expressed in all cells, coding for proteins having a housekeeping role. Contrary to this interpretation are the observations that most rare and moderately prevalent transcripts present in embryos are absent in adult tissues (Galau et al., 1976; Xin et al., 1982). The patterns of protein synthesis in several adult tissues are quite distinct from those in embryos as well (our unpublished observations). Similar to highly differentiated cells of other organisms, cells of specific adult sea urchin tissues show greatly enhanced synthesis and accumulation of a few polypeptides.

Expression of the Paternal Genome in Interspecies Hybrid Embryos

While it is well established that translation of maternal mRNA
accounts for most of the protein synthesis during the first few hours
of development, the contribution of transcription of the embryonic
genome to the mRNA population during early embryonic development has
been vague. One approach to this problem is to analyze the
expression of the paternal genome in interspecies hybrid embryos.
This approach, for instance, has demonstrated that H1 histone genes
are transcribed before the first cleavage and account for
translatable mRNA in early cleavage-stage embryos (Maxson and Egrie,
1980). We have analyzed the synthesis of paternal proteins in three
interspecies crosses (Tufaro and Brandhorst, 1982). These data are
summarized in Table 1. No paternal proteins were detected in early
cleaving embryos; it should be noted here that unlike the genes
coding for the early histone variants which are highly repeated
(Hentschel and Birnstiel, 1981), the genes coding for most of these
proteins are present in one or a few copies per haploid genome
(Goldberg et al., 1972; Davidson et al., 1975). Surprisingly, as
late as pluteus stage only a few specifically paternal proteins were
detected. This observation was confirmed for the S. purpuratus x L.
pictus cross using a radioactive complementary DNA probe transcribed
from polyadenylated, polysome-enriched gastrula RNA (Tufaro and
Brandhorst, 1982). The probe shares few sequences with RNA of S.
purpuratus. Comparison of the kinetics of hybridization of this
probe to excess RNA of L. pictus or hybrid gastrulae indicates that
most prevalent and moderately prevalent paternal sequences are
reduced or absent in hybrid embryos. Many rare paternal transcripts
are represented in hybrids. DNA-driven hybridization with this probe
indicates that the missing paternal RNA sequences are maintained in
the hybrid genome in normal amount.

There are two types of explanation for these observations. The
first is that expression of the paternal genome in hybrid embryos is
defective. For example, cytoplasmic factors of maternal origin
required for transcription or post-transcriptional processing may
exist and act in a species-specific manner. The observation that the
synthesis of paternal proteins is restricted in both of the
reciprocal crosses between S. purpuratus and S. droebachiensis
indicates that there is not a species-specific genomic or cytoplasmic
dominance. It also suggests that if cytoplasmic factors are involved
they have diverged rapidly in these two closely related species. It
is also possible that the maternal cytoplasm or genome might repress
the transcription of paternal genes as in the case of ribosomal genes
in hybrids of two species of Xenopus (Honjo and Reeder, 1973; Morgan
et al., 1982). It is also possible that the paternal genome is
rearranged in hybrid embryos.

The other type of explanation is that most prevalent mRNAs
translated in embryos are persistent maternal mRNAs synthesized

during oogenesis and not replaced by newly transcribed mRNA during embryonic development. In this case these paternal transcripts would not be expected to appear in normal amounts in the hybrid embryos. Analyses of the kinetics of accumulation of several transcripts which are prevalent in the egg using cloned DNA probes have indicated that these transcripts are rather stable and enter the cytoplasm slowly during embryonic development (Cabrera et al., 1983). The slow accumulation of new representatives of these sequences cannot account for their mass even late in development, indicating that these maternal transcripts are quite persistent. On the other hand, rare transcripts turn over rapidly, and newly transcribed rare mRNA has replaced maternal mRNA by gastrula stage. Paternal mRNAs of the complex (rare) class would thus be expected to be represented in hybrid embryos even if some prevalent maternal mRNAs are not normally replaced.

If some maternal mRNAs are persistent, some changes in gene expression during development may be the result of selective recruitment of these transcripts into polysomes. Much of the maternal RNA stored in the egg has a peculiar structure similar to that of unprocessed or incompletely processed nuclear precursors of mRNA. That is, much of the mass and complexity of maternal polyadenylated RNA includes interspersed, self-annealing, repeated sequences reminiscent of some introns (Costantini et al., 1980). Much of the egg RNA lacking poly(A) contains oligo(A) tracts characteristic of nuclear RNA but absent in polysomal mRNA (Duncan and Humphreys, 1981b). It is possible that these maternal RNAs give rise to actively translated mRNA after completion of post-transcriptional processing or modification (Davidson et al., 1982). Selective translational regulation of protein synthesis during embryonic development appears to operate in Ilyanassa (Brandhorst and Newrock, 1981; Collier and McCarthy, 1981). If maternal mRNAs are persistent and selectively modified for translation, they might be nonrandomly localized in the egg and/or segregated during cleavage into particular lineages (Brandhorst et al., 1983).

We are presently attempting to resolve these two possible explanations for the restricted expression of the paternal genome in hybrid embryos; both may apply to some extent. We have constructed a recombinant cDNA library complementary to mRNA prepared from L. pictus gastrulae and have identified several DNA clones corresponding to transcripts specific to L. pictus embryos which are considerably less prevalent (or absent) in hybrid embryos derived from S. purpuratus eggs fertilized with L. pictus sperm. These cloned cDNAs were used to characterize their respective transcripts in L. pictus embryos. Several of these transcripts are present in almost constant abundance in eggs and throughout embryonic development. These transcripts might correspond to persistent maternal RNAs, but alternatively, they might be replenished by new transcription. In several cases investigated thus far by hybridization of cloned DNA probes to blots of RNA separated electrophoretically by size, there is no detectible change in size of these transcripts during development. That is, there is no evidence that these transcripts are processed during development.

Other cDNA clones have been identified corresponding to transcripts which are barely detectible in hybrid embryos but accumulate during development of L. pictus due to new synthesis. The existence of this class of transcript indicates that at least some paternal genes normally expressed during embryonic development are hardly expressed in hybrid embryos. We are attempting to establish which events in gene expression account for the failure of these mRNAs to accumulate in hybrid embryos. The first step is to determine whether they are transcribed in hybrid embryos. Understanding the defect in expression of these paternal genes in hybrid embryos should provide valuable information about the interaction between nucleus and cytoplasm and the regulation of differential gene activity during embryonic development.

ACKNOWLEDGEMENTS

Our research was supported in part by grants from the National Science and Engineering Research Council of Canada and le Ministere de l'Education du Quebec. F.T. held a fellowship from McGill University. We thank Mary Bannet for technical assistance and help in preparation of this manuscript.

REFERENCES

Bedard, P.A. and Brandhorst, B.P. (1983). Patterns of protein synthesis and metabolism during sea urchin embryogenesis. Dev. Biol., in press.

Brandhorst, B.P. (1976). Two-dimensional gel patterns of protein synthesis before and after fertilization of sea urchin eggs. Dev. Biol. 52:310-317.

Brandhorst, B.P. (1980). Simultaneous synthesis, translation, and storage of mRNA, including histone mRNA in sea urchin eggs. Dev. Biol. 79:139-148.

Brandhorst, B.P., Bedard, P.A., and Tufaro, F. (1983). Patterns of protein metabolism and the role of maternal RNA in sea urchin embryos. In "Space, Time, and Pattern in Embryonic Development," W.R. Jeffery and R.A. Raff (eds.). Academic Press, New York, in press.

Brandhorst, B.P. and Humphreys, T. (1972). Stabilities of nuclear and messenger RNA molecules in sea urchin embryos. J. Cell Biol. 53:474-482.

Brandhorst, B.P. and Newrock, K.M. (1981). Post-transcriptional regulation of protein synthesis in Ilyanassa embryos and isolated polar lobes. Dev. Biol. 83:250-254.

Bruskin, A.M., Tyner, A.L., Wells, D.E., Showman, R.M., and Klein, W.H. (1981). Accumulation in embryogenesis of five mRNAs enriched in the ectoderm of the sea urchin pluteus. Dev. Biol. 87:308-318.

Bruskin, A.M., Bedard, P.A., Tyner, A.L., Showman, R.M., Brandhorst, B.P., and Klein, W.H. (1982). A family of proteins accumulating in ectoderm of sea urchin embryos specified by two related cDNA clones. Dev. Biol. 91:317-324.

Cabrera, C.V., Ellison, J.W., Moore, J.G., Britten, R.J., and Davidson, E.H. (1983). Regulation of cytoplasmic mRNA prevalence in sea urchin embryos: Rates of appearance and turnover for specific sequences. Submitted for publication.

Carson, D.D., Rossignol, D.P., and Lennarz, W.J. (1982). Induction of N-linked glycoprotein synthesis during gastrulation of sea urchin embryos. Cell Diff. 11:323-324.

Childs, G., Maxon, R., and Kedes, L.H. (1979). Histone gene expression during sea urchin embryogenesis: Isolation and characterization of early and late messenger RNAs of Strongylocentrotus purpuratus by gene-specific hybridization and template activity. Dev. Biol. 73:153-173.

Collier, R.J. and McCarthy, M. (1981). Regulation of polypeptide synthesis during early embryogenesis of Ilyanassa obsoleta. Differentiation 19:31-46.

Costantini, F., Britten, R.J., and Davidson, E.H. (1980). Message sequences and short repetitive sequences are interspersed in sea urchin egg poly(A)$^+$ RNAs. Nature (London) 287:111-117.

Crain, W.R., Durica, D.S., and Van Doren, K. (1981). Actin gene expression in developing sea urchin embryos. Mol. Cell Biol. 1:711-720.

Davidson, E.H., Hough-Evans, B.R., Klein, W.H., and Britten, R.J. (1975). Structural genes adjacent to interspersed repetitive DNA sequences. Cell 4:217-225.

Davidson, E.H., Hough-Evans, B.R., and Britten, R.J. (1982).
    Molecular biology of the sea urchin embryo. Science 217:17-26.

Duncan, R. and Humphreys, T. (1981a). Most sea urchin maternal mRNA
    sequences in every abundance class appear in both polyadenylated
    and nonpolyadenylated molecules. Dev. Biol. 88:201-210.

Duncan, R. and Humphreys, T. (1981b). Multiple oligo(A) tracts
    associated with inactive sea urchin maternal mRNA sequences.
    Dev. Biol. 88:211-219.

Durica, D.S. and Crain, W.R. (1982). Analysis of actin synthesis in
    early sea urchin development. Dev. Biol. 92:418-427.

Ernst, S.G., Hough-Evans, B.R., Britten, R.J., and Davidson, E.H.
    (1980). Limited complexity of the RNA in micromeres of 16-cell
    sea urchin embryos. Dev. Biol. 79:119-127.

Evans, T., Hunt, T., and Youngblood, J. (1982). On the role of
    maternal mRNA in sea urchins: Studies of a protein which
    appears to be destroyed at a particular point during each
    division cycle. Biol. Bull. 163:372.

Flytzanis, C.N., Brandhorst, B.P., Britten, R.J., and Davidson, E.H.
    (1982). Developmental patterns of cytoplasmic transcript
    prevalence in sea urchin embryos. Dev. Biol. 91:27-35.

Fregien, N., Dolecki, G.J., Mandel, M., and Humphreys, T. (1982).
    Molecular cloning of five individual stage- and tissue-specific
    messenger sequences from sea urchin pluteus embryos. Mol. Cell
    Biol., in press.

Galau, G.A., Britten, R.J., and Davidson, E.H. (1974). A
    measurement of the sequence complexity of polysomal messenger
    RNA in sea urchin embryos. Cell 2:9-20.

Galau, G.A., Klein, W.H., Davis, M.M., Wold, B.J., Britten, R.J., and
    Davidson, E.H. (1976). Structural gene sets active in embryos
    and adult tissues of the sea urchin. Cell 7:487-505.

Goldberg, R.B., Galau, G.A., Britten, R.J., and Davidson, E.H.
    (1972). Non-repetitive DNA sequence representation in sea
    urchin embryo messenger RNA. Proc. Natl. Acad. Sci. USA
    70:3516-3520.

Goustin, A.S. and Wilt, F.H. (1981). Protein synthesis,
    polyribosomes, and peptide elongation in early development of
    Strongylocentrotus purpuratus. Dev. Biol. 82:32-40.

Gross, P.R., Malkin, J.L., and Moyer, W.A. (1964). Templates for the first proteins of embryonic development. Proc. Natl. Acad. Sci. USA 51:407–414.

Grunstein, M. (1978). Hatching in the sea urchin Lythechinus pictus is accompanied by a shift in histone H4 gene activity. Proc. Natl. Acad. Sci. USA 75:4135–4139.

Harkey, M.A. and Whiteley, A.H. (1982a). The translational program during the differentiation of isolated primary mesenchyme cells. Cell Diff. 11:325–329.

Harkey, M.A. and Whiteley, A.H. (1982b). Cell-specific regulation of protein synthesis in the sea urchin gastrula: A two-dimensional electrophoretic study. Dev. Biol. 93:453–462.

Heifetz, A. and Lennarz, W.J. (1979). Biosynthesis of N-glycosidically linked glycoproteins during gastrulation of sea urchin embryos. J. Biol. Chem. 254:6119–6127.

Hentschel, C.C. and Birnstiel, M.L. (1981). The organization and expression of histone gene families. Cell 25:301–313.

Hieter, P.A., Hendricks, M.B., Hemminki, K., and Weinberg, E.S. (1979). Histone gene switch in the sea urchin embryo. Identification of late embryonic histone messenger ribonucleic acids and the control of their synthesis. Biochemistry 18:2707–2716.

Hinegardner, R.T. (1967). Echinoderms. In "Methods in Developmental Biology," F.H. Wilt and N.K. Wessells (eds.), pp. 139–155. Crowell Co., New York.

Honjo, T. and Reeder, R.H. (1973). Preferential transcription of Xenopus laevis ribosomal RNA in interspecies hybrids between Xenopus laevis and Xenopus mulleri. J. Mol. Biol. 80:217–228.

Hough-Evans, B.R., Wold, B.J., Ernst, S.G., Britten, R.J., and Davidson, E.H. (1977). Appearance and persistence of maternal RNA sequences in sea urchin development. Dev. Biol. 60:258–277.

Humphreys, T. (1969). Efficiency of translation of mRNA before and after fertilization in sea urchins. Dev. Biol. 20:435–458.

Humphreys, T. (1971). Measurements of messenger RNA entering polysomes upon fertilization of sea urchin eggs. Dev. Biol. 26:210–218.

Hutchins, R. and Brandhorst, B.P. (1979). Commitment to vegetalized development in sea urchin embryos: Failure to detect changes in patterns of protein synthesis. Wilhelm Roux' Archiv. 186:95-102.

Kuhn, O. and Wilt, F.H. (1981). Chromatin proteins of sea urchin embryos. Dev. Biol. 85:416-423.

Lasky, L.A., Lev, Z., Xin, J.H., Britten, R.J., and Davidson, E.H. (1980). Messenger RNA prevalence in sea urchin embryos measured with cloned cDNAs. Proc. Natl. Acad. Sci. USA 77:5317-5321.

Martindale, M.Q. and Brandhorst, B.P. (1982). The role of germinal vesicle and 1-methyladenine-induced changes in protein synthesis in Asterias oocytes. Biol. Bull. 163:374.

Maxson, Jr., R.E. and Egrie, J.C. (1980). Expression of maternal and paternal histone genes during early cleavage stages of the echinoderm hybrid Strongylocentrotus purpuratus x Lytechinus pictus. Dev. Biol. 74:335-342.

McClay, D.R. and Chambers, A.F. (1978). Identification of four classes of surface antigens appearing at gastrulation in sea urchin embryos. Dev. Biol. 63:179-186.

Merlino, G.T., Water, R.D., Moore, G.P., and Kleinsmith, L.J. (1981). Change in expression of the actin gene family during early sea urchin development. Dev. Biol. 85:505-508.

Morgan, G.T., Bakken, A.H., and Reeder, R.H. (1982). Transcription of Xenopus borealis rRNA genes in nuclei of Xenopus laevis oocytes. Dev. Biol. 93:471-477.

Newport, J. and Kirschner, M. (1982a). A major developmental transition in early Xenopus embryos: I. Characterization and timing of cellular changes at the midblastula stage. Cell 30:675-686.

Newport, J. and Kirschner, M. (1982b). A major developmental transition in early Xenopus embryos: II. Control of the onset of transcription. Cell 30:687-696.

Newrock, K.M., Cohen, L.H., Hendricks, M.B., Donelly, R.J., and Weinberg, E.S. (1978). Stage-specific mRNAs coding for subtypes of H2A and H2B histones in the sea urchin embryo. Cell 14:327-336.

O'Farrell, P.H. (1975). High resolution two-dimensional electrophoresis of proteins. J. Biol. Chem. 250:4007-4021.

Okazaki, K. (1975). Spicule formation by isolated micromeres of the sea urchin embryo. Amer. Zool. 15:567-581.

Raff, R.A. and Showman, R.M. (1983). Maternal messenger RNA: Quantitative, qualitative, and spatial control of its expression in embryos. In "The Biology of Fertilization," C.B. Metz and A. Monroy (eds.). Academic Press, New York, in press.

Rodgers, W.H. and Gross, P.R. (1978). Inhomogeneous distribution of egg RNA sequences in the early embryo. Cell 14:279-288.

Rosenthal, E.T., Brandhorst, B.P., and Ruderman, J.V. (1982). Translationally mediated changes in patterns of protein synthesis during maturation of starfish oocytes. Dev. Biol. 91:215-220.

Schneider, E.G., Nguyen, H.T., and Lennarz, W.J. (1978). The effect of tunicamycin, an inhibitor of protein glycosylation, on embryonic development of the sea urchin. J. Biol. Chem. 253:2348-2355.

Showman, R.M., Wells, D.E., Anstrom, J., Hursh, D.A., and Raff, R.A. (1982). Message-specific sequestration of maternal histone mRNA in the sea urchin egg. Proc. Natl. Acad. Sci. USA 79:5944-5947.

Tufaro, F. and Brandhorst, B. (1979). Similarity of proteins synthesized in isolated blastomeres of early sea urchin embryos. Dev. Biol. 72:390-397.

Tufaro, F. and Brandhorst, B.P. (1982). Restriction expression of paternal genes in sea urchin interspecies hybrids. Dev. Biol. 92:209-220.

Venezky, D.L., Angerer, L.M., and Angerer, R.C. (1981). Accumulation of histone repeat transcripts in sea urchin egg pronucleus. Cell 24:385-391.

Wells, D.E., Showman, R.M., Klein, W.H., and Raff, R.A. (1981). Delayed recruitment of maternal mRNA in sea urchin embryos. Nature 292:477-478.

Wilt, F.H. (1977). The dynamics of maternal poly(A)-containing mRNA in fertilized sea urchin eggs. Cell 11:673-681.

Xin, J-H., Brandhorst, B.P., Britten, R.J., and Davidson, E.H. (1982). Cloned embryo mRNAs not detectably expressed in adult sea urchin coelomocytes. Dev. Biol. 89:527-531.

TRANSLATIONAL REGULATION OF GENE EXPRESSION IN EARLY DEVELOPMENT

Joan V. Ruderman, Eric T. Rosenthal, and Terese Tansey[*]

ABSTRACT

Fertilization of Spisula oocytes causes a rapid change in the overall pattern of protein synthesis. This change occurs independently of any new mRNA transcription: it is controlled entirely at the translational level. cDNA clones complementary to several translationally regulated mRNAs have been isolated and used to directly investigate fertilization-triggered changes in mRNA activity and structure. Several mRNAs gain a long poly(A) tail right after fertilization, whereas others lose their poly(A) tails. In general, there is a good correlation between possession of a poly(A) tail and translational activity in vivo. Except for these changes in poly(A), these mRNAs show no significant structural alterations. This result strongly suggests that the inactivity of maternal mRNAs in the oocyte is not due to their being stored as translationally incompetent larger precursor forms. Evidence for a role of masking components is also discussed.

INTRODUCTION

Almost all embryos make use of a two-part strategy to meet the demands of cleavage and early development. Pools of structural proteins, enzymes, ribosomes and mRNAs are synthesized during oogenesis and stored in the mature oocyte. The activated oocyte and early embryo draw on these pools in an orderly temporal and spatial

---

[*]From the Program in Cell and Developmental Biology, and Department of Anatomy, Harvard Medical School, Boston, MA 02115.

61

way to produce cascades of metabolic changes and specific sets of
proteins and organelles. Almost all of these early changes are
independent of any ongoing nuclear transcription from the zygotic and
embryonic genome. As cleavage proceeds, the maternal mRNA and
protein pools are gradually depleted. At the same time, the numbers
of embryonic nuclei increase exponentially and their transcriptional
contributions become more significant. Finally, by the late blastula
or equivalent stage most of the original maternal contributions are
exhausted and further development is dependent on transcripts
produced by the embryo's genome. The exact details of the extent to
which the embryo relies on maternal vs. embryonic contributions and
the timing of this transition from one source to the other vary both
among individual organisms and according to the individual component.
However, in virtually all systems studied the immediate changes in
protein synthesis that accompany oocyte activation and fertilization
are due to changes in the translatability of the maternal mRNA pool.

From the few organisms that are accessible to studies of
oogenesis, we know that a large fraction of the oocyte's store of
maternal mRNA is synthesized very early in oogenesis. For example,
pre-vitellogenic Xenopus oocytes contain nearly their full complement
of poly(A)$^+$ RNA and mRNA sequence complexity (reviewed by Davidson,
1976). Studies of histone mRNAs and poly(A)$^+$ sequences complementary
to cloned cDNA probes show that these individual mRNAs follow the
same pattern of accumulation early in oogenesis (Golden et al., 1981;
van Dongen et al., 1981). In sea urchins, about half of the maternal
RNA sequence complexity is synthesized prior to the onset of
vitellogenesis (Hough-Evans et al., 1979). The remaining sequences
accumulate during the latter part of oogenesis. The bulk of the
maternal histone mRNA pool is made very late in oogenesis, between
the time of meiotic maturation and fertilization (Showman et al.,
1983; Angerer et al., 1983).

The maternal mRNA pool is phenomenologically a very special
class of mRNA. Some of these molecules (generally less than 1%), are
actively translated on polysomes in the oocyte, but the vast majority
of maternal mRNA sequences in the mature oocyte are translationally
inert. They are in some way "stored" for future use (reviewed by
Davidson, 1976; Raff and Showman, 1983). Fertilization (or meiotic
activation) triggers a rapid mobilization of some of these stored
mRNAs onto polysomes and, in some organisms, the reduction or
cessation of the translation of certain other previously active mRNAs
(Humphreys, 1971; Gross et al., 1973; Ballantine et al., 1979;
Rosenthal et al., 1980, 1982, 1983; Braude et al., 1979; Cascio and
Wassarman, 1982).

Despite much work in this area over the past 10 years, we
understand very little about the processes that control these
translational changes. The early experiments of Gross et al. (1973)
demonstrated that the sea urchin egg contains maternal mRNAs

(including histone mRNAs) that can be recovered after phenol extraction and are perfectly active in heterologous cell-free protein synthesizing systems. This result firmly established the idea that the mRNA itself was perfectly OK and that some other component in the egg was responsible for maternal mRNA unavailability. Several experiments supported the idea (originally put forward by Spirin in 1966 and expanded by Gross, Raff and colleagues) that maternal mRNAs are associated with proteins in messenger ribonucleoproteins (mRNPs), that these proteins "mask" the translational activity of most maternal mRNAs in the egg and that fertilization leads to an unmasking of mRNA which results in an increase in protein synthesis (Gross, 1967; Jenkins et al., 1978; Ilan and Ilan, 1978; Young and Raff, 1979). For example, Kaumeyer et al., (1978) isolated egg RNPs and found that this preparation was inactive in directing protein synthesis in an in vitro cell-free protein synthesizing system, whereas the phenol-extracted RNAs from the RNP preparation were active in vitro. However, not all workers in this field are in agreement on this point. Moon et al. (1982) carried out very similar experiments and found that their RNP preparations were just as active as the phenol-extracted RNAs isolated from the RNPs. Difficulties in precisely quantifying the mRNA content of various mRNP preparations and the template-activity of such diverse mRNA populations may contribute to the disparity of the two kinds of results. The use of cloned probes complementary to individual mRNAs should help to resolve this controversy.

In contrast to this idea of masked mRNA, some maternal mRNA sequences may be stored as larger, incompletely processed precursor RNAs. Costantini et al. (1980) found that 70% of the poly(A)$^+$ RNA complexity in the unfertilized sea urchin egg is present in very large RNA molecules that have several features of nuclear RNA. These RNAs appear to consist of interspersed, covalently linked transcripts of unique and repetitive sequences. Such molecules are not detected on polysomes in the embryo (Thomas et al., 1981). One interpretation of this finding is that these sequences could represent translationally inactive mRNA precursors that are processed after fertilization to produce active mRNA sequences. So far, there is only one example of a bona fide mRNA sequence of known coding assignment that fits this scheme. α-tubulin RNA sequences in the mature sea urchin egg are considerably larger than the usual ($\sim$ 2000 nucleotide) length found in embryos and adult tissues. These egg sequences are reduced to the usual size within 30 min of fertilization (Alexandraki and Ruderman, 1982, 1983). However, this phenomenon cannot explain the translational block on all stored mRNAs, since many abundant sea urchin maternal mRNAs are active in cell-free systems (Gross et al., 1973; Ruderman and Pardue, 1977; Jenkins et al., 1978; Ilan and Ilan, 1978). Also, at least one class of maternal mRNAs, those encoding the histones, do not undergo any detectable size change after fertilization (Lifton and Kedes, 1976; Wells et al., 1981).

Various other components of the translational machinery have been implicated in these translational switches. In urchins there is a 2-3 fold increase in the elongation rate, but this is too low to account for the overall 10-30 fold increase in the rate of protein synthesis (Brandis and Raff, 1978, 1979; Hille and Albers, 1979). Ballinger and Hunt (1981) found that fertilization causes a rapid and complete phosphorylation of one of the 40S ribosomal subunit proteins. However, careful comparisons of the kinetics of this phosphorylation with those of the rate increase, and of the representation of this protein in polysomal ribosomes vs. unrecruited ribosomes appear to rule out a direct and causal relationship (Ballinger, Doniach and Hunt, personal communication). Finally the experiments of Laskey et al. (1977) and of Richter and Smith (1981) show that globin mRNA injected into Xenopus oocytes competes on a 1:1 basis with endogenous polysomal mRNA and not on a 1:100 basis with the entire pool of maternal mRNA. Their results, as well as those of Lingrel and Woodland (1974) show that there is no spare translational capacity in the Xenopus oocyte and indicate that mRNa availability (i.e., mRNP masking) cannot be the sole factor regulating the rate of protein synthesis.

We have been investigating the molecular mechanisms of translational regulation in the marine mollusc Spisula. Oocytes and early embryos of this common clam contain mRNA populations that are indistinguishable by cell-free translation of their phenol-extracted mRNAs. When the oocytes are fertilized, there is a rapid and easily observed change in the pattern of protein synthesis. In this paper, we briefly review earlier results demonstrating that this change is regulated entirely at the translational level, and then present some recent experiments designed to get at the underlying mechanisms of this change. cDNA clones containing sequences complementary to several translationally regulated maternal mRNAs have been constructed and used to directly investigate changes in mRNA structure and activity during early development. Almost all of the Spisula maternal mRNAs examined so far show a change in polyadenylation at fertilization: some poly(A)-deficient oocyte mRNAs gain a long poly(A) tail immediately after fertilization, whereas other mRNAs became deadenylated during this same time. In many of these cases, there is a very close correlation between stage-specific adenylation and translational activity, but this does not seem to hold up for all maternal mRNAs examined. Our results also show that, except for the addition and removal of poly(A), fertilization does not cause any signficant changes in mRNA size, indicating that processing of larger translationally inactive precursors is not responsible for the activation of the particular mRNAs. Finally, we review evidence that implicates some other non-mRNA component in the regulation of stage-specific mRNA utilization.

METHODS

Culture and labeling of oocytes and embryos. Spisula
solidissima were collected from local waters by the Marine Resources
Staff at the Marine Biological Laboratory, Woods Hole, Massachusetts.
Gametes were prepared as described earlier (Rosenthal et al., 1980).
Oocytes and embryos to be labeled with radioactive amino acids were
pelleted gently and resuspended in 20 volumes of Millipore-filtered
sea water containing antibiotics. ^{35}S-methionine (New England
Nuclear, 600 Ci/mmole) was added to a final concentration of 0.2
mCi/ml, and oocytes and embryos were labeled in vivo for 20 min.
Samples were prepared for electrophoresis on polyacrylamide gels
containing SDS.

Preparation of 12K supernatants. Oocytes or embryos were washed
twice with cold calcium- and magnesium-free sea water and once with
homogenization buffer (0.3 M glycine, 70 mM potassium gluconate, 45
mM KCl, 2.3 mM $MgCl_2$, 1 mM DTT, 40 mM HEPES, pH 6.9) (Winkler and
Steinhardt, 1981). Cells were resuspended in 3 vol of homogenization
buffer and broken in a Dounce homogenizer. Homogenates were spun at
12,000 x g for 20 min. Aliquots of these "12K" supernatants were
frozen in liquid nitrogen and stored at -80°C.

Polysome gradients. 12K supernatants were diluted with 5 vol of
gradient buffer (0.5 M KCl, 6 mM $MgCl_2$, 0.5 mM DTT, 1 mM EDTA, 10 mM
HEPES, pH 7.0). Half of each sample was treated with 30 mM EDTA to
release mRNA from polysomes (Young and Raff, 1979). 1 ml of 12K
supernatant was layered onto a 15-40% sucrose (w/v) 11 ml gradient
made up in gradient buffer, and gradients were centrifuged in an SW41
rotor at 40,000 rpm for 70 min. Fractions were collected and RNA was
extracted. RNA from each gradient fraction was dissolved in an equal
volume of water, usually 30 µl, and stored at -80°C.

Cell-free translation. RNA samples were translated in a
mRNA-dependent rabbit reticulocyte lysate (Pelham and Jackson, 1976).
Total RNA was translated at a final concentration of 600 µg/ml. When
polysomal gradient RNA fractions were translated, 2 µl of each RNA
sample was added to 8 µl reticulocyte lysate. When the translational
specificites of unextracted Spisula oocyte or embryo 12K supernatant
were assayed, 1 µl of unextracted 12K supernatant was diluted with 1
µl of water and then added to 8 µl of reticulocyte lysate. All
incubations were carried out for 1 hr at 30°C. Labeled protein
samples were analyzed on SDS-polyacrylamide slab gels.

Isolation of cloned cDNAs complementary to translationally
regulated mRNA. Double-stranded cDNAs complementary to poly(A)$^{+}$ RNA
from 2-cell or 18 hr embryos were synthesized and inserted into the
Pst I site of the plasmid pBR322 by the GC-tailing method. Bacterial
clones transformed with recombinant plasmids were isolated and
recombinant plasmids DNAs from selected bacterial colonies were

prepared from bacterial lysates (Roychoudhury et al., 1976; Villa-Komaroff et al., 1978; Grunstein and Hogness, 1975). The cDNA sequences carried by individual clones were determined by mRNA hybrid-selected translation (Ricciardi et al., 1979; Alexandraki and Ruderman, 1981) of the complementary mRNA from total 2-cell or 18 hr embryo RNA followed by translation in vitro and identification of the translation products on 1-D gel electrophoresis. Details of this aspect of the work are presented elsewhere (Rosenthal et al., 1983).

RNA blots. RNA samples were electrophoresed on 1% agarose gels containing formaldehyde (Rave et al., 1979), transferred to nitrocellulose and hybridized with ^{32}P-labeled plasmid DNA probes (Thomas, 1980). Plasmid DNAs were labeled in vitro with ^{32}P by nick-translation (Maniatis et al., 1975). Total RNA was fractionated on oligo(dT)-cellulose (Aviv and Leder, 1972) and, in some cases, on poly(U)-Sepharose (Pharmacia).

RNAse H removal of poly(A) tails. Unfractionated 12K supernatant RNAs were hybridized with oligo(dT) and the double-stranded portions of the hybrids were digested with RNAse H (Bethesda Research Laboratories) (Vournakis et al., 1975). Nuclease-resistant RNAs were then electrophoresed on an agarose gel, blotted to nitrocellulose and hybridized to ^{32}P-labelled cloned cDNA probes as above.

RESULTS

The Pattern of Protein Synthesis Changes Within Ten Minutes of Fertilization

Fertilization of the Spisula oocyte triggers breakdown of the germinal vesicle (10 min post-fertilization), first (40 min) and second (50 min) meiotic divisions. First cleavage occurs 70 min after fertilization and is unequal, producing a small AB and a large CD blastomere which have different developmental fates (Ruderman et al., 1983). Subsequent cleavages occur without any significant change in the mass of the embryo. About 5 hr after fertilization, the 60-cell embryo gastrulates and develops specialized ciliated regions. The pyramidal trochophore larva takes shape between 9 and 12 hr. At 12 hr the veliger larva beings to form: shell is secreted, muscle and nerve cells are produced and the gut becomes regionalized. At 20 hr, the clam-shaped veliger larva contains many differentiated organs, including a highly ciliated velum. This larval form persists for several days until metamorphosis occurs. During the first 24 hr of development, there are many changes in the patterns of protein synthesis (Figure 1). The earliest of these occurs between 2 and 10 min of fertilization and is regulated entirely at the translational level. As development continues and the embryo's store of maternal mRNA declines, new transcription makes

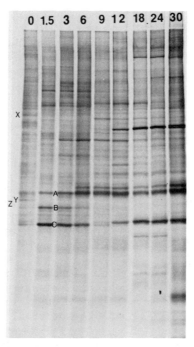

Figure 1: Autoradiogram of proteins synthesized in vivo by oocytes and embryos.

Oocytes (0) and embryos (1.5-30 hr post-fertilization) were labeled in vivo with ^{35}S-methionine. The labeled proteins were analyzed by electrophoresis on a 10% polyacrylamide-SDS slab gel followed by autoradiography.

increasingly important contributions to the total mRNA pool and to the pattern of protein synthesis (Tansey and Ruderman, 1983). In this paper, only the early changes in protein synthesis will be considered.

The Early Change in the Pattern of Protein Synthesis is Controlled Entirely at the Translational Level

Within 10 min of fertilization or artificial activation, synthesis of many oocyte-specific proteins ceases and the zygote begins to make new sets of proteins (Rosenthal et al., 1980). For example, the proteins marked X, Y and Z (Figure 1) are synthesized by the oocyte, but their synthesis is not detectable in zygotes. In contrast, proteins A, B, and C are synthesized at very low levels, or not at all, in oocytes whereas they are prominently labeled in 10-min embryos. Comparisons of the in vivo patterns with the in vitro

translation products programmed by oocyte and zygote RNAs in a
reticulocyte lysate (Figure 2) reveal that this abrupt change in the
kinds of proteins being made occurs in the absence of any detectable
changes in the total maternal mRNA pool. Phenol-extracted oocyte and
zygote RNAs appear to contain the same mRNA sets including the
sequences for proteins A, B, C, X, Y and Z.

The experiment shown in Figure 3 demonstrates that different
subsets of this maternal mRNA pool are found on polysomes before and
after fertilization. 12,000 x g supernatants of oocyte and zygote
homogenates were fractionated on sucrose gradients into polysomal and
post-ribosomal superantant fractions. RNA was extracted from these
fractions and translated in vitro. Figure 3 shows that a very
specific subset of oocyte mRNAs are engaged on polysomes and that
fertilization causes a dramatic redistribution of sequences between
the translated, polysomal compartment and the non-translated,
post-ribosomal supernatant compartment.

Analyses of the translational status of individual mRNAs by this
method are hindered by two problems. First, when the polypeptides
encoded in vitro by two or more mRNAs co-migrate, it is impossible to
conclude anything about the behavior of their respective mRNAs.
Secondly, since the translation of maternal mRNA is quantitatively
and qualitatively regulated in vivo and we do not understand the
molecular mechanisms responsible for this, seemingly straightforward
interpretations of in vitro translation assays could be wrong.
Therefore, we have used a second mRNA assay method, hybridization to
cloned cDNA probes, that does not depend on in vitro template
activity. Recombinant DNA clones consisting of pBR322 and cDNAs
complementary to translationally regulated maternal mRNAs were
isolated from three cDNA libraries (Rosenthal et al., 1983; Tansey
and Ruderman, 1983; Rosenthal and Ruderman, unpublished). Among the
clones used in the following experiments were those carrying
sequences for protein A (clone 1T55), protein C (clone 1T43), and
α-tubulin (clone 3V4). Several other clones that showed
hybridization to RNA on Northern blots of 2-cell RNA, but did not
selectively hybridize detectable amounts of template-active RNA, were
also used in this study.

These clones were first used to monitor the changes in the
translational utilization of several individual mRNAs at
fertilization. Oocyte and embryo homogenates were fractionated on
polysome gradients. RNAs were extracted from each of the gradient
fractions, electrophoresed on denaturing agarose gels, blotted to
nitrocellulose and hybridized with individual ^{32}P-labeled cloned
cDNAs. The hybrdization pattern obtained with clone 1T55 (protein A)
shown in Figure 4 reveals that none of mRNA A is present on polysomes
in the oocyte and that all of this mRNA moves onto polysomes right
after fertilization. The mRNA for protein C, as well as several
other, behaves similarly. Another pattern of maternal mRNA

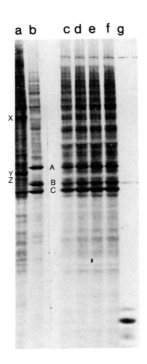

Figure 2: Comparison of proteins synthesized in vivo by oocytes and
embryos with those programmed in vitro by oocyte and embryo RNA.
Proteins were labeled with ^{35}S-methionine.

(a) Oocyte proteins synthesized in vivo; (b) 2-cell embryo proteins
synthesized in vivo. Lanes c-f, in vitro translation products
encoded by: (c) oocyte RNA; (d) embryo RNA, 15 min
post-fertilization; (e) embryo RNA, 50 min post-fertilization; (f)
2-cell embryo RNA; (g) endogenous incorporation (no RNA added).
From Rosenthal et al. (1980), Cell 20:487. Reproduced with
permission of MIT Press, Cambridge.

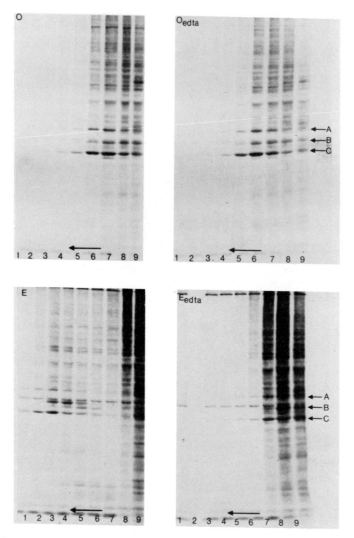

Figure 3: In vitro translation products labeled with ^{35}S-methionine
and directed by RNA extracted from sucrose gradient fractions of
oocyte and embryo 12K supernatants.

Top panels: Translation products encoded by total oocyte RNA.
Bottom panels: protein encoded by total embryo RNA. Translation
products directed by RNAs isolated from sucrose gradient
fractionations of 12K supernatants from oocytes (O) and embryos
(E), and of EDTA-treated 12K supernatants from oocytes (O$_{EDTA}$) and
embryos (E$_{EDTA}$). Arrow indicates direction of sedimentation. 60S
subunits are found in fractions 7 and 8; 40S subunits are found in
fractions 8 and 9.

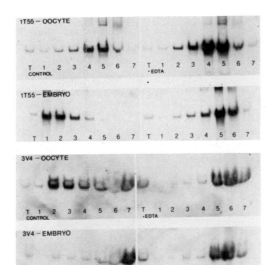

<u>Figure</u> <u>4</u> (left): Hybridization of ³²P-labeled 1T55 DNA and 3V4 DNA to Northern blots of RNA from polysome gradient fractions.

12K supernatants from oocytes and embryos were fractionated on sucrose gradients as described in Methods. In set of lanes marked +EDTA, the 12K supernatants were treated with EDTA prior to fractionation in order to release mRNA from polysomes. Direction of sedimentation is from right to left. An equal portion of the corresponding 12K supernatant was run in each lane marked "T."

recruitment is exemplified by 6T21 mRNA: these mRNAs (like A and C) are scarcely translated, if at all, in the oocyte but (unlike A and C) fertilization causes only a partial recruitment of these sequences (Rosenthal et al., 1983). A third kind of translational change is shown by α-tubulin mRNA (clone 3V4, Figure 4) and several other sequences of unknown coding assignment: a portion of these mRNAs are engaged on polysomes in the oocyte and fertilization results in their abrupt release from polysomes. Finally, a few of the cloned maternal mRNA sequences show no changes in the extent of translation after fertilization.

<u>Sequence-specific Changes in Adenylation are the Only Significant</u>
<u>Changes in Maternal mRNA Sites After Fertilization</u>

As one way of testing for mRNA precursor-product processing at fertilization, the sizes of the oocyte and zygote versions of several translationally regulated mRNAs were compared on "Northern blots." Equal amounts of oocyte and zygote total cellular RNA were electrophoresed in adjacent lanes of a denaturing formaldehyde-agarose gel, transferred to nitrocellulose by blotting, and hybridized with ^{32}P-labeled cloned cDNAs. Figure 5 shows the results obtained for two different sequences that become active after fertilization, mRNAs A (clone 1T55) and C (clone 1T43). Neither RNA sequence shows any evidence of being processed to a smaller size after fertilization. In fact, the zygote mRNAs are slightly <u>larger</u> than their oocyte counterparts, indicating that something has been added to these mRNAs after fertilization. Several other maternal mRNAs also show this slight (50-100 nucleotide) increase in length following fertilization. In contrast, α-tubulin mRNA (which is active in the oocyte and inactive in the zygote) becomes slightly shorter after fertilization.

Two kinds of experiments show that these length changes are due to the addition or removal of poly(A) tails at fertilization. First, total oocyte and zygote RNAs were fractionated by oligo(dT)-cellulose chromatography into unbound, poly(A)-deficient RNA and bound, poly(A)$^+$ RNA. The translation products encoded by these RNA fractions are shown in Figure 6. Whereas the <u>in vitro</u> translation products of total oocyte and zygote RNA appear identical, a very different picture is seen when the poly(A)$^-$ and poly(A)$^+$ products are compared. Fertilization causes changes in the abilities of many maternal mRNAs to bind to oligo(dT)-cellulose. The most dramatic examples include the mRNA for proteins A, B and C. The oocyte versions of these three mRNAs are poly(A)-deficient, whereas the embryo sequences are poly(A)$^+$. In contrast, other mRNAs (such as the one encoding the cell-free translation product marked by an asterisk) is poly(A)$^+$ in the oocyte and poly(A)$^-$ in the zygote. These changes occur right after fertilization, just at the time when these mRNAs move onto polysomes. This result shows that the mRNAs for proteins A, B, and C are poly(A)-deficient in the oocyte and become adenylated after fertilization, whereas other mRNAs become deadenylated during the same interval. This finding is supported by the experiment shown in Figure 7. Northern blots of total, poly(A)$^-$ and poly(A)$^+$ RNA fractions of oocytes and embryos were hybridized with several different ^{32}P-labeled cloned cDNAs. The hybridization pattern obtained with 1T55 (protein A) shows that essentially all of this mRNA is poly(A)-deficient in the oocyte and becomes poly(A)$^+$ after fertilization. Several other mRNAs, including mRNA C, show the same switch. In contrast, hybridizations with the α-tubulin probe (3V4) show that α-tubulin mRNA is poly(A)$^+$ before fertilization and poly(A)$^-$ after.

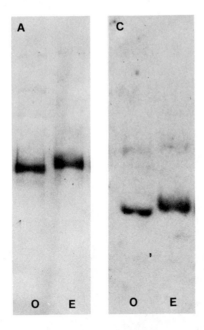

Figure 5: The amounts and sizes of mRNAs for proteins A (1T55) and C (1T43) before and after fertilization.

3 µg of total cellular RNA from oocytes and embryos was run in adjacent lanes on an agarose-formaldehyde gel, blotted to nitrocelulose, and hybridized with ^{32}P-labeled 1T55 (MRNA A) or 1T43 (mRNA C).

A second kind of experiment demonstrates that these additions and removals of poly(A) account for most or all of the mRNA size changes that happen at fertilization. When the sizes of the oocyte and embryo versions of individual mRNAs are compared after their poly(A) tails have been removed by RNAse treatment, they are indistinguishable. An example of this experiment using clone 1T55 (protein A) as the hybridization probe is shown in Figure 8. These and other results (Rosenthal et al., 1983) demonstrate that, except

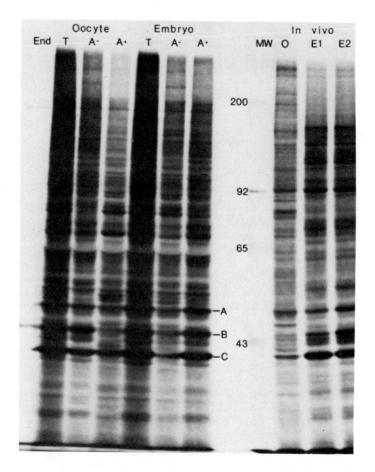

Figure 6: Translation products encoded by [T] total unfractionated RNA; [-] poly(A)⁻ RNA; and [+] poly(A)⁺ RNA from oocytes and embryos.

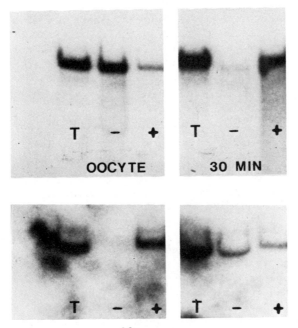

Figure 7: Hybridization of ^{32}P-cloned DNA probes 1T55 (protein A) and 3V4 (-tubulin) to total poly(A)$^-$ RNA and poly(A)$^+$ RNA from oocytes and 30 min post-fertilization embryos.

RNAs were electrophoresed on an agarose gel, blotted to nitrocellulose, and hybridized with ^{32}P-labeled cloned DNAs.

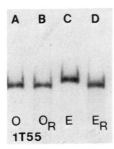

Figure 8: Comparison of the sizes of mRNA for protein A in oocytes and embryos.

RNAs were isolated from oocytes (O) and 2-cell embryos (E). In the lanes O_R and E_R the poly(A) tails were removed using kRNAse prior to electrophoresis. Samples were electrophoresed on an agarose gel, blotted to nitrocellulose, and hybridized with ^{32}P-labeled clone 1T55 DNA.

for the addition or removal of poly(A), there are no significant
changes in mRNA sizes that would indicate extensive processing of
larger precursors after fertilization. Of course, more subtle
alterations such as the removal of just a few nucleotides or the
modification of 5' cap structures cannot be ruled out in these
experiments.

The Relationship Between Adenylation and Translational Activity of Spisula Maternal mRNAs Is Good But Not Perfect

    In many of the cases examined so far, the correlation between
the adenylation and translation status of individual mRNAs in vivo is
intriguingly close. For example, several mRNAs such as A, B, and C
are poly(A)$^-$ in the oocyte where they are translationally repressed,
and these same mRNAs become polyadenylated at the time as they become
translationally active in the zygote. Other mRNAs, such as
α-tubulin, are both poly(A)$^+$ and active in the oocyte and become
deadenylated and inactivated after fertilization. However, not all
mRNAs fit this pattern. A few mRNAs are found in both the poly(A)$^+$
and poly(A)$^-$ fractions in the oocyte but show translational activity
at this stage (Rosenthal et al., 1983; Rosenthal and Ruderman,
unpublished).

Stage-specific mRNA Translation is Mimicked in a Mixed Cell-Free System

    Oocytes and embryos contain the same sets of mRNAs and these
mRNAs are equally translatable in vitro when the usual reticulocyte
cell-free lysate is used. Except for adenylation and deadenylations,
no obvious changes in mRNA size occur at fertilization, suggesting
that major processing events are not responsible for the
translational activation and repression of specific maternal mRNAs.
If this is true, then some other component must play an important
role in this translational switch. Earlier experiments attempted to
reproduce the stage-specific translation of individual mRNAs in a
cell-free system. Crude 12K supernatants were prepared from Spisula
oocytes or embryos homogenized in a buffer (0.3 M glycine, 70 mM
potassium gluconate, 45 mM KCl, 2.3 mM MgCl$_2$, 1 mM DTT, 40 mM HEPES,
pH 6.9) developed by Winkler and Steinhardt (1981) for a sea urchin
egg cell-free system. These 12K supernatants by themselves do not
initiate protein synthesis very efficiently. However, when the crude
12K supernatants are mixed with "helper" mRNA-dependent reticulocyte
lysate, this mixed cell-free system now initiates and synthesizes
protein rather actively (Figure 9). Furthermore, the translation
products encoded by the oocyte-reticulocyte mix closely resemble
those made by the oocyte in vivo, whereas the embryo-reticulocyte mix
yields a pattern of protein synthesis that is like the embryo pattern
in vivo. Thus, the initiations that occur in these mixed cell-free
systems retain their stage-specificity. Clearly, some translational
cues are present in these crude unextracted 12K supernatants and are

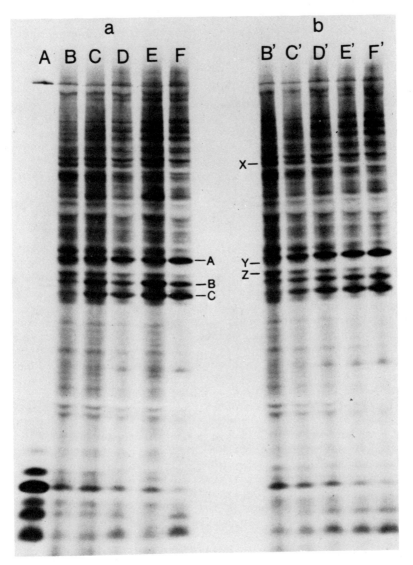

Figure 9: Results of mixing oocyte and embryo 12K supernatants
before translation in reticulocyte lysate.

(a) Proteins synthesized in the heterologous cell-free translation
system when unextracted Spisula oocyte 12K supernatants are mixed
with unextracted Spisula 2-cell embryo 12K supernatants and
translated in reticulocyte lysate. Patterns of protein synthesis
directed by reticulocyte lysate and: (A) endogenous
(homogenization buffer added instead of 12K supernatant); (B) 100%
oocyte 12K supernatant; (C) 25% embryo 12K supernatant--75% oocyte

lost during the phenol extraction of RNA from them. Little is known
about the nature of the "information" that is responsible for this
stage-specific, sequence-specific translational discrimination. When
oocyte and embryo 12K supernatants are mixed in varying proportions
prior to adding them to the reticulocyte lysate, the pattern of
protein synthesis simply resembles the sum of the input oocyte and
embryo patterns (Figure 9). This kind of result suggests that there
is no excess of positive or negative diffusible regulatory molecules.
One interpretation is that these mRNAs in vivo are complexed with
proteins as RNP particles in vivo and that certain of these proteins
can modulate translational activity in different ways under different
physiological circumstances. Another possibility is that large pH or
ionic differences in oocyte and zygote cytoplasms promote differences
in mRNA secondary structure that determine whether or not an mRNA is
available for ribosome binding.

DISCUSSION

     Fertilization of Spisula oocytes causes a rapid change in the
pattern of protein synthesis that is regulated entirely at the
translational level. Within 10 min of fertilization, many of the
mRNAs that were actively engaged in protein synthesis in oocytes are
released from polysomes and become translationally inactive in the
zygote. Examples include mRNAs X, Y, Z and α-tubulin. During this
same interval other sets of previously stored maternal mRNAs enter
polysomes and become translationally active. Some of these mRNAs,
such as A and C, are completely mobilized onto polysomes, whereas in
other cases only a fraction of the stored sequence is recruited.

     The molecular mechanisms that control these switches are
unknown. It seems unlikely that large changes in the primary
structure of these mRNAs play a role in this phenomenon in Spisula.
First, except for various adenylations and deadenylations, the mRNA
sizes do not change detectably at fertilization. However, small
changes in mRNA sizes, such as the removal of a few nucleotides or
the modification of a 5' cap, would not have been seen in our

Figure 9 (continued):

  12K supernatant; (D) 50% embryo 12K supernatant—50% oocyte 12K
  supernatant; (E) 75% embryo 12K supernatant—25% oocyte 12K
  supernatant; (F) 100% embryo 12K supernatant. (b) For comparison
  with (a), the heterologous cell-free trnaslation products shown in
  lanes B and F were mixed in comparable ratios after the translation
  reaction was terminated by addition of SDS gel sample buffer.
  Lanes B'-F' correspond to the same ratios as lanes B-F in (a).
  From Rosenthal et al. (1980), Cell 20:487. Reproduced with
  permission of MIT Press, Cambridge.

experiments. Second, the oocyte and zygote versions of these mRNAs are, after phenol extraction, equally translatable in the reticulocyte cell-free system.

An important caveat to this second line of reasoning is that at least one change in mRNA structure--namely adenylation--does occur in response to fertilization but this change has no detectable effect on the behavior of these mRNAs in vitro. The poly(A)$^+$ and poly(A)$^-$ versions of several mRNAs show very different translational activities in vivo but are translated equally well in the reticulocyte lysate. Thus it seems quite possible that other modifications of mRNA primary structure may be occurring and that such modifications could be recognized and obeyed in the oocyte/zygote milieu, but that these differences are simply ignored by a less discriminatory, heterologous in vitro system. Only direct sequence comparisons of maternal mRNAs from oocytes and zygotes will properly resolve this issue.

The role of polyadenylation in translational activation and repression is similarly murky. Several translationally regulated mRNAs undergo a change in polyadenylation soon after fertilization. In some cases, such as mRNAs A, B and C, translational activation of these sequences occurs during the same 10-min interval that they become polyadenylated. In other cases, such as α-tubulin and certain other sequences of unknown coding identity, loss of their translational activity after fertilization is accompanied by deadenylation. However, other mRNAs do not show such a close coupling between adenylation and activity. Most of these comparisons have been carried out using oligo(dT)-cellulose, which fractionates the RNAs into those sequences containing fewer than about 6-10 A residues and those containing longer A tracts. Perhaps some other length differential is used for these translational decisions. If that were so, then the oligo(dT)-cellulose fractionation test would sometimes, but not always, show a correlation between possession of a poly(A) tail and translational activity in vivo (Rosenthal et al., 1983).

Changes in adenylation of maternal mRNAs have been seen in other species as well. In the starfish, for example, hormonal activation of the oocyte sets off a translational change that closely resembles that in Spisula; several of the activated mRNAs are converted from poly(A)$^-$ to poly(A)$^+$ forms (Rosenthal et al., 1983; Rosenthal, Brandhorst and Ruderman, unpublished observations). In Xenopus, actin mRNAs become deadenylated after oocyte maturation (Colot and Rosbash, 1982), and cease translation roughly at the same time (Ballantine et al., 1979), but no direct correlation has been made yet. In contrast, deadenylation of the Xenopus oocyte's poly(A)$^+$ histone mRNA pool at maturation occurs about the same time as the selective recruitment of histone mRNA onto polysomes (Adamson, and Woodland, 1977; Ruderman et al., 1979). In this last case, the

relationship between adenylation and translation is the opposite of
the one usually seen. However, the failure of all mRNAs to follow a
single rule may simply mean that there is more than one rule.

In Dictyostelium, the transition from a vegetative to an
aggregation ("developmental") program is accompanied by the
deadenylation of actin mRNA and the loss of that mRNA from the
polysomes (Palatnik et al., 1982). In one of the few relatively
direct tests for an involvement of poly(A) in translation, Jacobson
and Favreau (1983) found that free poly(A) tracts inhibit the rate of
translation in vitro of many poly(A)$^+$ RNAs whereas the translation
of poly(A)$^+$ mRNAs is not affected by the poly(A). These authors
suggest that the binding of an mRNA's poly(A) tail to the poly(A)
binding protein in some way causes an enhancement of that mRNA's
initiation rate. Indeed, early experiments by Jeffery and Brawerman
(1975) indicated that the poly(A) tails of mouse sarcoma mRNAs
interact closely with other parts of the mRNAs. It should be
possible to test for such associations in Spisula maternal mRNAs now
that cloned probes have been made.

In certain systems, a case can be made that when the overall
rate of protein synthesis changes, intrinsic differential efficiences
of mRNA initiation will lead to changes in the extent of mRNA
competition and therefore to different patterns of protein synthesis
(Lodish, 1974; Godefroy-Coburn and Thach, 1982). It seems very
unlikely that ths is the explanation for the changes in maternal
mRNAs translation. In Spisula, the overall rate of protein synthesis
increases only 2-4X after fertilization, yet the pattern of protein
synthesis changes very dramatically. In fact, certain mRNAs are not
translated at all in the oocyte and are completely recruited onto
polysomes after fertilization. These same general features appear to
be true for starfish (Rosenthal et al., 1982) and possibly for
Xenopus (Adamson and Woodland, 1977; Ballantine et al., 1979; Meuler
and Malacinski, this volume). In sea urchins, the rate of protein
synthesis goes up sharply, 10-50 fold depending on the species (Epel,
1967), but the pattern of proteins synthesized by the egg and the
zygote are for the most part indistinguishable (Brandhorst, 1976).
Thus, the most dramatic, sequence-specific changes in mRNA
translation happen in the face of a very insignificant rate change
(Spisula), and the fewest changes occur despite extremely large rate
increases after fertilization (sea urchin).

Finally, the role of sequence-specific sequestration and
stage-specific release of maternal mRNAs should be considered. This
sequestration could be gross, with certain mRNAs being held within
organelles, or could be executed at the molecular level by masking
proteins that bind to an mRNA and prevent its translation. Both
cytological fractionation and in situ hybridization experiments show
that maternal histone mRNAs in sea urchin eggs are restricted to the
egg pronucleus and remain there until just before first cleavage,

when they then exit to the cytoplasm (Showman et al., 1982, and this volume; Angerer et al., this volume). This explanation, however, probably does not hold for the translationally regulated mRNAs in starfish and Spisula. Cytoplasmic fractionation experiments reveal no enrichment for any of these RNAs in the oocyte germinal vesicle (Martindale and Brandhorst, 1982; Rosenthal and Ruderman, unpublished).

Molecular sequestration still remains an open possibility. Whereas phenol-extracted mRNAs from oocytes and zygotes are equally active in vitro in the reticulocyte lysate, crude unextracted oocyte and zygote homogenates added to reticulocyte lysate program the oocyte- and zygote-specific patterns of protein synthesis, respectively. Homogenate mixing experiments suggest that there are no diffusible regulators, a finding also consistent with the idea of masking proteins. However, this is not the only interpretation. Perhaps other RNA-binding molecules arbitrate these translational decisions. New insights will require new experimental approaches.

ACKNOWLEDGEMENTS

This work was supported by NSF Grant PCM 7923046 and N.I.H. Research Career Development Award K04 HD00349 to J.V.R.

REFERENCES

Adamson, E.D. and Woodland, H.R. (1977). Changes in the rate of histone synthesis during oocyte maturation and very early development of Xenopus laevis. Develop. Biol. 57:136-149.

Alexandraki, D. and Ruderman, J.V. (1983). Changes in tubulin mRNA sequences during sea urchin early development. In preparation.

Alexandraki, D. and Ruderman, J.V. (1981). Sequence heterogeneity, multiplicity and genomic organization of α- and β-tubulin genes in sea urchins. Mol. Cell. Biol. 1:1125-1137.

Alexandraki, D. and Ruderman, J.V. (1982). Organization and expression of the tubulin gene families in the sea urchin. J. Submic. Cytol., in press.

Angerer, R.C., Hughes, K.J., DeLeon, D.V., Lynn, D.A., and Angerer, L.M. (1983). [This volume].

Aviv, H. and Leder, P. (1972). Purification of biologically active globin messenger RNA by chromatography on oligothymidylic acid-cellulose. Proc. Natl. Acad. Sci. USA 69:1408-1413.

Ballantine, J.E.M., Woodland, H.R., and Sturgess, E.A. (1979). Changes in protein synthesis during development of Xenopus laevis. J. Embryol. Exp. Morphol. 51:135-153.

Ballinger, D.G. and Hunt, T. (1981). Fertilization of sea urchin eggs is accompanied by 40S ribosomal subunit phosphorylation. Dev. Biol. 87:277-285.

Brandhorst, B.P. (1976). Two-dimensional gel patterns of protein synthesis before and after fertilization of sea urchin eggs. Develop. Biol. 52:310-317.

Brandis, J.W. and Raff, R.A. (1978). Translation of oogenetic mRNA in sea urchin eggs and early embryos. Demonstration of a change in translational efficiency following fertilization. Dev. Biol. 67:99-113.

Brandis, J.W. and Raff, R.A. (1979). Elevation of protein synthesis is a complex response to fertilization. Nature 278:476-469.

Braude, P., Pelham, H., Flach, T., and Lobatto, R. (1979). Post-transcriptional control in the early mouse embryo. Nature 282:102-105.

Cascio, H.V. and Wassarman, P.M. (1982). Program of early development in the mammal: Post-transcriptional control of a class of proteins synthesized by mouse oocytes and early embryos. Dev. Biol. 89:397-408.

Colot, H.V. and Rosbash, M. (1982). Behavior of individual maternal pA$^+$ RNAs during embryogenesis of Xenopus laevis. Dev. Biol. 94:79-86.

Costantini, F.E., Britten, R.J., and Davidson, E.H. (1980). Message sequences and short repetitive sequences are interspersed in sea urchin egg poly(A) RNAs. Nature (London) 287:111-117.

Davidson, E.A. (1976). "Gene Activity in Early Development." Academic Press, New York.

Epel, D. (1967). Protein synthesis in sea urchin eggs: A "late" response to fertilization. Proc. Natl. Acad. Sci. USA 57:899-906.

Godefroy-Coburn, T. and Thach, R.E. (1982). The role of mRNA competition in regulating translation. IV. Kinetic Model. J. Biol. Chem. 256:11762-11773.

Golden, L., Schafer, U., and Rosbash, M. (1981). Accumulation of individual pA RNAs during oogenesis of Xenopus laevis. Cell 22:835-844.

Gross, P.R. (1967). The control of protein synthesis in embryonic development and differentiation. Curr. Topics Dev. Biol. 2:1-29.

Gross, K.W., Jacobs-Lorena, M., Baglioni, C., and Gross, P.R. (1973). Cell-free translation of maternal messenger RNA from sea urchin eggs. P.N.A.S. 70:2614-2518.

Grunstein, M. and Hogness, D.S. (1975). Colony hybridization: a method for the isolation of clone DNAs that contain a specific gene. Proc. Natl. Acad. Sci. USA 75:5544.

Hille, M.B. and Albers, A.A. (1979). Efficiency of protein synthesis after fertilization of sea urchin eggs. Nature 278:469-471.

Hough-Evans, B.R., Ernst, S.G., Britten, R.J., and Davidson, E.H. (1979). RNA complexity in developing sea urchin oocytes. Dev. Biol. 69:258-269.

Humphreys, T. (1971). Measurements of messenger RNA entering polysomes upon fertilization of sea urchin eggs. Dev. Biol. 26:201-208.

Ilan, J. and Ilan, J. (1978). Translation of maternal message ribonucleoprotein particles from sea urchin in a cell-free system and product analysis. Dev. Biol. 66:375-385.

Jacobson, A. and Favreau, M. (1983). Possible involvement of poly(A) in protein synthesis. Submitted for publication.

Jeffery, W.R. and Brawerman, G. (1975). Association of the polyadenylate segment of messenger RNA with other polynucleotide sequences in mouse sarcoma 180 polyribosomes, Biochemistry 14:3445.

Jenkins, N.A., Kaumeyer, J.F., Young, E.M., and Raff, R.A. (1978). A test for masked message: The template activity of messenger of ribonucleprotein particles isolated from sea urchin eggs. Dev. Biol. 63:279-298.

Kaumeyer, J.F., Jenkins, N.A., and Raff, R.A. (1978). Messenger ribonucleoprotein particles in unfertilized sea urchin eggs. Dev. Biol. 63:266-278.

Laskey, R.A., Mills, A.D., Gurdon, J.R., and Partington, G.A.
    (1977). Protein synthesis in oocytes of Xenopus laevis is not
    regulated by the supply of messenger RNA. Cell 11:345-351.

Lifton, R.P. and Kedes, L.H. (1976). Size and sequence homology of
    masked maternal and embryonic histone mRNAs. Dev. Biol.
    48:47-55.

Lingrel, J.B. and Woodland, H.R. (1974). Initiation does not limit
    the rate of globin synthesis in message-injected Xenopus
    oocytes. Eur. J. Biochem. 47:47-56.

Lodish, H.F. (1974). Model for the regulation of mRNA translation
    applied to haemoglobin synthesis. Nature 251:385-388.

Maniatis, R., Jeffrey, A., and van de Sande, H. (1975). Chain
    length determination of small double and single-stranded DNA
    molecules by polyacryalmide gel electrophoresis. Biochem.
    14:3787-3794.

Martindale, M. and Brandhorst, B. (1982). Biol. Bull. (Abstracts)
    in press.

Moon, R.T., Danilcheck, M.V., and Hille, M.B. (1982). An assessment
    of the masked message hypothesis: sea urchin egg messenger
    ribonucleoprotein complexes are efficient templates for in vitro
    protein synthesis. Dev. Biol. 93:389-403.

Palatnik, C.M., Wilkins, C., and Jacobson, A. (1982). Submitted for
    publication.

Pelham, H.R.B. and Jackson, R.J. (1976). An efficient
    mRNA-dependent translation system from reticulocyte lysates.
    Eur. J. Biochem. 67:247-256.

Raff, R.A. and Showman, R.M. (1983). Maternal mRNA: quantitative,
    qualitative and spatial control of its expression in embryos.
    In "The Biology of Fertilization," C.B. Metz and A. Monroy
    (eds.), in press.

Rave, M., Crkvenjokov, R., and Boedtker, H. (1979). Identification
    of procollagen mRNAs transferred to diazabenzyloxymethyl paper
    from formaldehyde agarose gels. Nucleic Acids Res.
    6:3559-3567.

Ricciardi, R.P., Miller, J.S., and Roberts, B.E. (1979).
    Purification and mapping of specific mRNAs by
    hybridization-selection and cell-free translation. Proc. Natl.
    Acad. Sci. USA 76:4927-4931.

Richter, J.D. and Smith, L.D. (1981). Differential capacity for translation and lack of competition between mRNAs that segregate to free- and membrane-bound polysomes. Cell 27:183-191.

Rosenthal, E., Hunt, T., and Ruderman, J.V. (1980). Selective translation of mRNA controls the pattern of protein synthesis during early development of the surf claim Spisula solidissima embryos. Cell 20:487-496.

Rosenthal, E.T., Brandhorst, B.P., and Ruderman, J.V. (1982). Translationally mediated changes in patterns of protein synthesis during maturation of starfish oocytes. Dev. Biol. 91:215-220.

Rosenthal, E.T., Tansey, T.R., and Ruderman, J.V. (1983). Sequence-specific adenylations and deodynylations accompany changes in the translation of maternal mRNA after fertilization of Spisula oocytes. J. Mol. Biol., in press.

Roychoudhury, R., Jay, E., and Wu, R. (1976). Terminal labeling and addition of homopolymer tracts to duplex DNA fragments by teminal deoxynucleotidyl transferase. Nucleic Acids Res. 3:101-116.

Ruderman, J.V. and Pardue, M.L. (1977). Analysis of mRNA in echinoderm and amphibian early development. Dev. Biol. 60:48-68.

Ruderman, J.V., Tansey, T.R., Rosenthal, E.T., Hunt, T., and Cheney, C.M. (1983). Spatial and temporal aspects of gene expression during Spisula embryogenesis. In "Embryos: Time, Space, and Pattern," R.A. Raff and W. Jeffery (eds.). A.R. Liss, New York, in press.

Ruderman, J.V., Woodland, H.R., and Sturgess, E.R. (1979). Modulations of histone messenger RNA during early development of Xenopus laevis. Dev. Biol. 71:71-82.

Showman, R.M., Wells, D.E., Anstrom, J., Hursh, D.A., and Raff, R.A. (1982). Message-specific sequestration of maternal histone mRNA in the sea urchin egg. Proc. Natl. Acad. Sci. 79:5944-5947.

Showman, R., Wells, D., Anstrom, J.A., Hursh, D.A., Leaf, D.S., and Raff, R.A. (1983). [This volume].

Spirin, A.S. (1966). On "masked" forms of messenger RNA in early embryogenesis and in other differentiating systems. Curr. Top. Dev. Biol. 1:1-38.

Tansey, T.R. and Ruderman, J.V. (1983). Changing patterns of
    protein synthesis in Spisula embryos are controlled by changes
    in both mRNAs levels and translatable utilisation. Submitted.

Thomas, T.L., Posakony, J.W., Anderson, D.M., Britten, R.J., and
    Davidson, E.R. (1981). Molecular structure of maternal mRNA.
    Chromosoma 84:319-335.

Thomas, P. (1980). Hybridization of denatured RNA and small DNA
    fragments transferred to nitrocellulose. Proc. Natl. Acad. Sci.
    77:5201-5205.

Van Dongen, W., Zoal, R., Moorman, A., and Destree, O. (1981).
    Quantitation of the accumulation of histone messenger RNA during
    oogenesis in Xenopus laevis. Dev. Biol. 86:303-314.

Villa-Komaroff, L., Efstratiadis, A., Broome, S., Lomedico, P.,
    Tizard, R., Naber, S.P., Chick, W.S., and Gilbert, W. (1978).
    A bacterial clone synthesizing·proinsulin. Proc. Natl. Acad.
    Sci. USA 76:3683-3687.

Vournakis, J.H., Efstratiadis, A., and Kafatos, F.C. (1975).
    Electrophoretic patterns of deadenylated chorian and globin
    mRNAs. P.N.A.S. 72:2959-2963.

Wells, D.E., Showman, R.M., Klein, W.H., and Raff, R.A. (1981).
    Delayed recruitment of maternal histone mRNA in sea urchin
    embryos. Nature (London) 292:477-479.

Winkler, M.M. and Steinhardt, R.A. (1981). Activation of protein
    synthesis in a sea urchin cell-free system. Dev. Biol.
    84:432-439.

Young, E.M. and Raff, R.A. (1979). Messenger ribonucleoprotein
    particles in developing sea urchin embryos. Dev. Biol.
    72:24-40.

mRNA DISTRIBUTIONS IN SEA URCHIN EMBRYOS

Robert C. Angerer, Kathleen J. Hughes, Donna V. DeLeon,
David A. Lynn, and Lynne M. Angerer[*]

ABSTRACT

We have developed techniques for localization of individual mRNA
species in sections of sea urchin embryos using the hybridization of
radioactively labeled RNA probes to tissue sections in situ. The
method yields grain densities proportional to the concentration of
target mRNAs and is sufficiently sensitive to allow the localization
of most moderately abundant mRNAs. The major features of the method
are briefly discussed, and several examples of its use are reviewed.
We show that maternal histone mRNA is largely sequestered in
pronuclei of unfertilized eggs and released at nuclear membrane
breakdown of first cell division. We discuss several events in
differentiation of cells of the dorsal ectoderm, including expression
of a cell-type specific set of messenger RNAs.

INTRODUCTION

During the past decade, the process of development of the sea
urchin embryo from egg to pluteus larva has been extensively analyzed
using the increasingly powerful methods of molecular genetics. While
the urchin embryo has been used to examine general questions of gene
expression (Davidson, 1976; Davidson et al., 1982) as well as the
structure and control of individual genes (for a recent review see
Davidson et al., 1982), perhaps the major emphasis has been placed on
investigations of the requirements for differential gene expression

---

[*]From the Department of Biology, University of Rochester,
Rochester, NY 14618.

during the events of determination and differentiation. Analysis of the sequence complexity of RNA in unfertilized eggs showed that the embryo has an initial store of maternal RNA with a high potential information content, about 3×10^7 NT (Hough-Evans et al., 1977). Since the number average length of maternal RNA is about 3000 NT (Costantini et al., 1980), this sequence complexity indicates that this RNA contains transcripts from about 15,000 different genes. At least 75% of these sequences appears to be associated with potential messenger RNA since that fraction of the complexity can be recovered in polysomal RNA isolated from 16-cell embryos (Ernst et al., 1980). Similar analyses of polysomal mRNAs at progressively later stages of development revealed the somewhat surprising fact that the sequence complexity declines by a factor of 3 during development during pluteus (Galau et al., 1976). Furthermore, most sequences present at later stages are also present in the unfertilized egg. While these results initially implied a similar three-fold decrease in the number of different mRNAs required as embryogenesis progresses, recent analyses of maternal RNA structure indicate that this simple interpretation is not strictly correct. For example, Costantini et al. (1980) have shown that maternal RNA has several characteristics of incompletely processed nuclear transcripts, including large size and a high content of interspersed repetitive sequences. Thus, part of the overall decrease in polysomal RNA complexity during development may be attributable to degradation and/or processing of maternal transcripts and their replacement with (or persistence of) shorter mRNAs at later stages. The unusual structure of maternal RNA and our lack of knowledge about its exact relationship to functional messenger RNA currently make it impossible to do accurate bookkeeping on the number of different mRNAs expressed in whole embryos at different developmental stages. Nevertheless, it is clear that the increase in morphological complexity during embryogenesis is not accompanied by a large increase in the number of different mRNAs present in the embryo.

The above analyses of mRNA complexity are largely weighted by a diverse set of messages present in the whole embryos at low concentration; i.e., several thousand copies per embryo (or several copies per cell at pluteus). Modulations in the population of more highly abundant mRNAs, comprising 10-20% of the different species, have been examined by several laboratories using different techniques. Analysis of proteins translated from moderately prevalent RNAs by Brandhorst (1976) indicated that most of these are synthesized in the embryo throughout development, although quantitative modulations occur. (The reader is referred to the chapter by Brandhorst in this volume for a review of this extensive data.) These findings agree with measurements of prevalence of individual messages using recombinant cDNA probes which showed that only 10-20% of moderately abundant mRNAs undergo relatively large changes in whole embryo abundance during the course of development (Flytzanis et al., 1982). It is clear, however, that measurements of

whole embryo prevalence may average out greater modulations in the
concentration of any specific mRNA in different cell lineages. Thus,
a complete picture of patterns of gene expression requires knowledge
of the spatial distribution of individual mRNAs in the embryo at
different developmental stages.

There are several approaches for examining cell lineage
specificity of expression of individual genes or sets of genes. Cell
separation methods have been developed for purification of micromeres
from 16-cell embryos (Hynes and Gross, 1970), for isolation of the
descendents of the micromeres, the primary mesenchyme cells (Ernst,
personal communication), and for separation of the three primary germ
layers (ectoderm, mesenchyme, and endoderm) from gastrulae and plutei
(McClay and Marchase, 1979). Such fractionations have allowed
analyses of proteins synthesized (Tufaro and Brandhorst, 1979), and
of RNA complexity (Rodgers and Gross, 1978; Ernst et al., 1980) in
subregions of the embryo, and identification and isolation of
individual mRNAs enriched in ectoderm (Bruskin et al., 1981; and see
below). In general, these techniques are somewhat limited since 1)
they allow separation of cell types only after their developmental
fates are at least partly determined and 2) in most cases individual
cell types cannot be isolated. Two in situ methods offer the
possibility of identifying precise sites of expression of individual
genes: antibodies may be used to localize specific proteins, and in
situ hybridization may identify cells containing specific RNAs.

Application of cell separation and molecular localization
techniques individually and in combination allow examination of
spatial patterns of expression of individual genes throughout
development. Such analyses will help answer a variety of general and
specific questions. One set of general questions deals with the true
diversity of gene expression among different embryonic cell lineages.
Determination of the distributions of a number of individual
moderately abundant messenger RNAs at a variety of developmental
stages will provide estimates of the fraction of these mRNAs that are
specific to (or enriched in) one or more lineages. This information
is necessary to determine the number of different lineages in the
embryo and the times at which each becomes differentiated at the
level of mRNA content. In essence, this is an attempt to construct a
"fate map" of the embryo in molecular terms and to relate timing of
differential gene expression to that of overt differentiation. More
specific questions focus on the function of individual genes or gene
families. For example, during cleavage the sea urchin embryo
expresses a large set of tandemly repeated genes coding for "early"
histone variants, while at later stages, synthesis of these mRNAs is
discontinued and transcription of a different set of "late" variant
mRNAs is initiated (for reviews see Kedes, 1979; and Hentschel and
Birnsteil, 1981). Analysis of the timing of this switch in different
cell lineages may provide clues to cellular events which regulate it
and help relate this switch to other events of differentiation in the

embryo. For example, it is important to know know whether all cells of the embryo make the switch at the same time or if different lineages make the switch at different times, perhaps correlated with changes in the rate of cell division and/or onset of expression of lineage-specific mRNAs. For other individual mRNAs (and the corresponding proteins), identification of cell types expressing them may be helpful in defining function. For example, the sea urchin genome contains a rather large number of genes encoding actins (Scheller et al., 1981), at least some of which are expressed at specific developmental stages (Crain et al., 1981). Correlation of the expression of individual genes with specific cell types and morphogenetic events would help establish the function of different members of this gene family and the proteins they encode. Finally, efforts have already begun in several laboratories to identify the cell type-specific mRNAs and proteins whose functions may then be investigated. One example of such an analysis is discussed below.

During the past several years, our laboratory has developed and improved techniques for detecting individual mRNAs in sea urchin embryos by hybridization of pure nucleic acid sequences to tissue sections in situ. In the following sections the current status of this technology is discussed briefly and several of its initial applications are reviewed to illustrate the kind of new information that can be obtained.

In Situ Hybridization Technique

Detection of RNA transcripts by in situ hybridization requires compromises among several conflicting requirements: The method of fixation should provide good histological preservation so that different cell types can be recognized and must also provide high retention of target RNAs in sections throughout the hybridization procedure. At the same time, target RNAs must be (or subsequently be made to be) accessible to hybridization probes, and they must not be chemically modified to the extent that they will not hybridize with the required specificity. For an initial approach to this problem, we hybridized radioactively labeled poly U to polyadenylated RNAs in sectioned embryos (Angerer and Angerer, 1981). The results of this study showed that for embryos of the sea urchin (Strongylocentrotus purpuratus), fixation in glutaraldehyde provides good preservation of morphology and high retention of RNAs. However, the fraction of target poly A sequences accessible to hybridization with poly U is quite low in glutaraldehyde-fixed material. A large increase in hybridization efficiency was achieved by partial deproteinization with proteinase K, as originally suggested by Brahic and Haase (1978) for tissue fixed with ethanol-acetic acid.

The final factor in improving the sensitivity was the nature of the probe used. Two different types of probe were considered, and the choice between them was not initially clear. Symmetric probes (denatured, double-stranded; for example, DNA labeled by nick translation) offered the potential advantage that multi-stranded hyperpolymers might form on target RNAs in situ, increasing the amount of probe bound, and hence the sensitivity of the assay. However, results in several laboratories, including ours (Brahic and Haase, 1978; Angerer and Angerer, 1981), indicated that relatively short fragments of probe penetrate into sections more efficiently. Thus, duplexes formed by probe self-reassociation in solution might not subsequently hybridize efficiently in situ. We have recently compared directly the hybridization efficiency of symmetric and asymmetric (mRNA-complementary strand only) probes (Hughes et al., manuscript in preparation). For this analysis we hybridized RNA probes representing the 6.3 kb early histone gene repeat sequence from pCO2 (Overton and Weinberg, 1978), encoding each of the five histone mRNAs, to sections of 12-hr embryos containing high concentrations of early variant histone mRNAs. After hybridization under identical conditions, the signal obtained with the asymmetric probe is about eight-fold higher at apparent saturation than that obtained with the symmetric probe.

Our current methodology for in situ situ hybridization is described in Lynn et al. (1983) and summarized briefly in Table 1. As probes, we used ^{3}H-labeled RNAs, transcribed in vitro from recombinant DNA templates as prepared by inserting sea urchin DNA sequences in a transcription vector, RVIIΔ7. This vector was constructed by E. Butler and P. Little in the laboratory of T. Maniatis. The sea urchin DNA sequences are inserted within a multiple cloning site just downstream from a Sp6 Salmonella phage promoter. The templates are usually truncated by restriction digestion at a site just downstream from the insert, and asymmetric RNA run-off transcripts are synthesized in vitro using purified Sp6 RNA polymerase and ^{3}H-NTPs (Butler and Chamberlin, 1982). The advantage of using truncated templates is that the resulting probes contain little contaminating vector sequence. This decreases the total concentration of probe required to achieve saturation of target RNAs in situ and consequently decreases the level of nonspecific binding of probe to sections. After synthesis, the template DNA is removed by DNase I digestion, and the probe size is reduced to 100-150 bases by controlled alkaline hydrolysis.

We have characterized many aspects of in situ hybridization reactions using the early histone repeat probe as a model system (Hughes et al., manuscript in preparation). Our thermal denaturation studies in solution showed that RNA-RNA duplexes are considerably more stable than DNA-DNA or DNA-RNA duplexes. This difference is intensified in buffers containing formamide, as might be expected from previously published reports on relative stability of DNA duplexes and DNA-RNA hybrids in such buffers (Casey and Davidson, 1977). Comparison of the thermal stability of hybrids formed in situ

Table 1. Detection of mRNAs by In Situ Hybridization

	Procedure

Tissue Preparation

a) Fix in 1% glutaraldehyde.
b) Embed in paraffin.
c) Section 5 μ nominal thickness.

Glutaraldehyde provides superior morphological preservation and increased retention of cellular RNAs (Angerer and Angerer, 1981). Fixed tissues can be stored several years in 70% EtOH or in paraffin without noticeable changes in hybridization.

Prehybridization Treatments

a) Deparaffinize and hydrate.
b) Proteinase K, 1 μg/ml, 37°C, 30 min.
c) Acetic anhydride, 0.25%, 25°C, 10 min.

Protease K increases the signal at least ten-fold, presumably by removing proteins from the target RNAs; it does not decrease RNA retention at 1 μg/ml (Angerer and Angerer, 1981). Acetic anhydride blocks amino groups and reduces electrostatic binding of probe to sections (Hayashi et al., 1978).

Hybridization

^{3}H Asymmetric RNA Probe, ∼ 150 b.
Buffer: 50% formamide; 0.3 M NaCl; 10 mM Tris; 1 mM EDTA, pH 8.0; 0.2% each Ficoll, BSA, and PVP; 500 μg tRNA/ml; 10% dextran sulfate; 500 μg poly(A)/ml for inserts cloned by AT tailing.
Hybridize 45–50°C overnight.

Asymmetric probes provide much higher signals than denatured duplex probes (Hughes et al., manuscript in preparation). Ficoll, BSA, PVP, and tRNA are competitors for nonspecific binding.

Procedure

Posthybridization Washes

a) 4 x SSC, 25°C, brief.
b) 20 μg RNase A/ml in 0.5 M NaCl, 10 mM Tris,
   1 mM EDTA, pH 8.0; 30 min, 37°C.
c) 2 x SSC and 0.1 x ssc, 25°C, 30 min each.

Dextran sulfate increases the hybridization rate (Lederman et al., 1981). The incubation temperature is 25°C below the Tm for RNA-RNA duplexes formed in situ (Hughes et al., manuscript in preparation).

Autoradiography

a) Dip NTB-2 emulsion in 0.3 M ammonium acetate.
b) Incubate in moist chamber 3 hr.
c) Expose at 4°C in vacuum dessicator.
d) Develop D-19, stop 2% HAc, fix; all at 15°C.
e) Stain, permount.

Incubation in a moist chamber reduces latent grains in the emulsion. Staining protocols which use acidic pH may artificially decrease grain densities (Rogers, 1979; our unpublished observations).

showed that they have Tm's about 5°C lower than RNA-RNA hybrids formed with the same probes in solution. We believe that this difference is primarily due to shorter length of hybrids formed in situ, although we cannot exclude the possibility that part of this decrease is due to chemical alteration of some nucleotides during the fixation procedure. The temperature for optimum rate of hybridization in situ is approximately 25°C below the Tm of the duplexes formed, as has been observed for solution hybridizations (Wetmur and Davidson, 1968). We compared thermal stabilities of hybrids formed between the early variant histone probe and either homologous early mRNAs or divergent late variant mRNAs, both in solution and after in situ hybridization to early (12-hr) and late (pluteus) stages. The same difference in Tm (13°C) was observed for homologous vs. heterologous hybrids in both cases, showing that the stringency of hybridization can be adjusted for in situ reactions.

Hybridizations in situ yield signals which increase in proportion to probe concentration until the target RNAs are saturated (Hughes et al., manuscript in preparation). Estimates of hybridization efficiency based on the known content of histone mRNAs per embryo indicate that about 90% of total target sequence is both retained in sections and hybridized with probe at saturation (see Angerer and Angerer, 1981, for a discussion of efficiency estimates as well as the uncertainties involved). Thus, we believe that little further increase in efficiency can be obtained by modifications of the present hybridization technique. In two cases we have shown that grain densities over sections are proportional to relative contents of target RNA sequences measured by standard biochemical techniques (Angerer and Angerer, 1981; Hughes et al., manuscript in preparation). As illustrated below, the sensitivity of the assay is sufficient to allow detection of many moderately abundant mRNAs.

Histone mRNAs in Pronuclei of Unfertilized Eggs

The unfertilized egg of S. purpuratus contains an early variant histone message comprising several percent of the stored maternal mRNA. Early studies by Skoultchi and Gross (1973) indicated that essentially all translatable histone mRNA was cytoplasmic since it could be recovered from 40s RNP particles in cytoplasmic supernatants. In situ hybridization of nick translated pCO2 DNA, containing one copy of each early histone gene as well as interspersed spacer sequence, to unfertilized eggs yielded a surprising result: grain densities over pronuclei were about 50 fold higher than over surrounding cytoplasm (Venezky et al., 1981). Although signals over cytoplasm were quite low, after correction for the large ratio of cytoplasmic to nuclear volumes we concluded that as much as 15% of target RNA sequences were contained in pronuclei.

Because in these initial studies the probe included both mRNA-coding and spacer sequences, we could not specifically identify

the sequences contained in nuclear transcripts. There were at least
four possibilities for the nature of these transcripts: 1) Sequences
with homology to the spacers might occur on a variety of other
nuclear RNAs. 2) The nuclear transcripts might result from "read
through" of long stretches of the tandemly repeated histone genes. A
similar phenomenon, resulting from apparent failure of transcript
termination, has been well documented in oocytes of Notophthalmus
(Diaz et al., 1981). 3) The transcripts could consist of discrete
precursors to individual histone mRNAs whose abnormally slow
processing results in their accumulation in pronuclei. 4) The
nuclear RNAs could be mature histone mRNAs which accumulate because
of a deficiency in transport.

To distinguish among such possibilities we analyzed the sequence
content of pronuclear RNAs in detail (DeLeon et al., manuscript in
preparation). Since it has not been possible to isolate intact
pronuclei containing high concentrations of histone gene transcripts
(Skoultchi and Gross, 1973; Showman et al., 1981; our unpublished
observations), we used in situ hybridization for this analysis. The
coding sequences for all five histone mRNAs are contained on the same
strand of the repeat (see Kedes, 1979, for review). The histone
repeat from pCO2 was transferred into the RVIIΔ7 vector in both
insert orientations, and separate RNA probes representing the coding
strand and mRNA strand were synthesized. The coding strand probe
hybridized intensely to pronuclei, while the mRNA strand gave signals
not distinguishable from nonspecific background. Next we examined
the hybridization of nine separate subclones encompassing the early
repeat. All probes containing sequences coding for any of the five
histone mRNAs gave signals over pronuclei, whereas three probes
containing only spacer sequence failed to hybridize. Comparison of
the hybridization of two partially overlapping pairs of probes
showed, in the case of H2A and H3 mRNAs, that the pronuclear
transcripts contain only mRNA sequence and no detectable spacer
sequence. Finally, we compared the thermal stabilities of hybrids to
histone mRNAs in two-cell or 12-hr embryo cytoplasm. The Tm's of
these hybrids were indistinguishable.

The simplest interpretation of these results is that the
pronuclear transcripts are mature histone mRNAs. This conclusion
receives further support from observations on the fate of these
pronuclear RNAs after fertilization and from improved estimates of
the fraction of total maternal histone mRNA contained in pronuclei.
The striking change in distribution of these RNAs after fertilization
is illustrated in Fig. 1. Sections of eggs fixed at 10-min intervals
after fertilization were hybridized with the early repeat probe. By
two-cell stage a large increase in grain density over cytoplasm is
observed (Fig. 1), whereas there is little hybridization to two-cell
nuclei (Venezky et al, 1981). We quantitated this increase in
cytoplasmic grain density by measurements over randomly selected
embryos at each time point (Fig. la, ●-●). We also estimated the

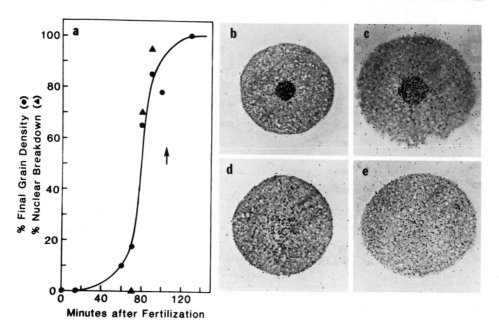

Figure 1: Change in the localization of histone mRNAs after
fertilization. Sea urchin eggs were fixed before fertilization and
at successive 10-min intervals after fertilization. Sections were
hybridized with the early histone repeat RNA probe. a) After
autoradiography grain densities were determined over the cytoplasm
of randomly selected sections and normalized to the plateau value
at late 2-cell stage (●). (There is no significant increase in
total histone mRNA content between fertilization and late 2-cell
state.) The percent nuclear breakdown was estimated from the
number of intact nuclei in random sections at each time point (Δ).
The arrow indicates the time at which 50% of the embryos cleaved.
b-e) Representative sections of embryos fixed at 70, 80, 80, and 90
min, respectively.

time of nuclear membrane breakdown during first cleavage by counting
the number of intact nuclei in randomly selected sections. The
percent of embryos which had undergone nuclear breakdown at three
time points is also indicated in Fig. 1a (Δ-Δ). At the level of
resolution of a few minutes, the increase in cytoplasmic signal is
coincident with nuclear breakdown. Since there is no increase in
histone mRNA content during this period (Maxson and Wilt, 1981; Wells
et al., 1981), we believe these data demonstrate that a very large
fraction of maternal histone mRNA is sequestered in pronuclei and
subsequently released to the cytoplasm at nuclear breakdown. This
suggestion is rather strikingly supported by observations of patterns
of in situ hybridization observed over different sections fixed at

different points in this transition, as shown in Fig. 1b-e. Such
patterns strongly suggest the release of histone mRNAs from pronuclei
and their rather rapid equilibration throughout the cytoplasm.

The redistribution of histone mRNAs at first cleavage made
possible the more accurate quantitation of the fraction of histone
mRNA contained in pronuclei. Since we do not currently have reliable
estimates of the efficiency of hybridization to nuclear RNAs, we
calculated this fraction from the difference in cytoplasmic grain
densities for unfertilized eggs and two-cell embryos. These
measurements for probes representing both the whole early repeat unit
and several individual histone mRNAs indicated that at least 90-95%
of maternal early variant histone mRNA is restricted to pronuclei in
unfertilized eggs (DeLeon et al., manuscript in preparation).
Northern blot analysis of RNA isolated from whole eggs shows that
these histone mRNAs are indistinguishable from early variant histone
mRNAs in the cytoplasm of embryos at later stages (DeLeon et al.,
manuscript in preparation), reinforcing the conclusion that the
pronuclear transcripts are mature histone mRNAs.

Raff and his co-workers have reached similar conclusions by
Northern blotting analysis of RNAs contained in nucleated and
enucleated egg fragments (Showman et al., 1982) and more recently by
analysis of RNAs contained in nuclei fractionated from ethanol-fixed
eggs. (See the chapter by Showman et al. in this volume. The reader
is also referred to that chapter for a discussion of the possible
biological significance of this phenomenon.)

Studies on the localization of histone mRNAs in unfertilized
eggs have thus revealed an unexpected and unusual situation where
expression of these genes is interrupted specifically at the step of
mRNA transport from the nucleus. Two obvious potential control
points are the transport apparatus of the pronucleus and structure of
the histone mRNAs. Analysis by in situ hybridization indicates that
pronuclei do not accumulate polyadenylated RNAs (Angerer and Angerer,
1981), and labeling studies by Brandhorst (1980) have shown that at
least some RNAs synthesized in unfertilized eggs are both transported
to the cytoplasm and loaded on polysomes. To date, pronuclear
accumulation has been demonstrated only for early variant histone
mRNAs. It may be significant that these mRNAs probably do not have
longer precursors (Mauron et al., 1982) and are neither spliced nor
polyadenylated. Thus, histone mRNAs may utilize a
processing/transport pathway which is different from that of most
other messages and which is inactive in unfertilized eggs. A more
unusual alternative, which currently has no direct experimental
support, is that early variant histone mRNAs lack sequence signals
required for transport from nuclei. It is interesting, though
perhaps only coincidental, that the early variant histone mRNAs are
synthesized in post-fertilization embryos only at stages where
continuous cell divisions occur, each accompanied by breakdown of

nuclear membranes. Whatever the mechanism, the phenomenon may not be restricted to unfertilized eggs and zygotes, but persists as late as 16-cell stage. The rate of early variant mRNA synthesis is very low for the first few hours of development and increases markedly at about 16-cell stage (Maxson and Wilt, 1981). In situ hybridizations of the whole repeat probe to two-, four-, and eight-cell embryos reveals very low signals over nuclei, consistent with this lack of synthesis. As shown in Fig. 2, at 16-cell stage the signals are several fold higher over nuclei than over surrounding cytoplasm. While these signals are much lower than those measured for pronuclei, it is likely that a true steady state concentration would not have been achieved in the few minutes since a high rate of synthesis was initiated.

Differentiation of the Dorsal Ectoderm

We have collaborated with the group of Dr. William Klein in determining the localization of a set of mRNAs designated Spec 1 mRNAs. (The reader is referred to the chapter by Klein et al. in this volume for a discussion of the properties of this small family of genes and their RNA transcripts.) pSpec 1 was one of several cDNA clones identified in Klein's laboratory as representing a set of mRNAs enriched in isolated ectoderm. Spec 1 mRNAs are essentially absent from maternal RNA and increase greatly in concentration as the result of new embryonic transcription beginning at blastula around 20 hr of development (Bruskin et al., 1981).

The pSpec 1 cDNA insert was transferred to the transcription vector in both orientations and coding (signal), and mRNA strand (control) probes were synthesized (Lynn et al., 1983). Examples of hybridizations to sections of 68-hr plutei are shown in Fig. 3. Measurements of grain densities over different defined regions of these sections indicate that hybridization above background levels is observed only over cells of dorsal ectoderm. There is no detectable signal over gut or over two other regions of ectoderm--the growing arms and the ventral area around the mouth. The labeled dorsal ectoderm appears to consist largely of a morphologically uniform cell type. Estimates made from stained whole mounts indicate that there are about 400 of these cells at this stage. Measurements of the absolute number of Spec 1 mRNA transcripts per pluteus by standard solution titration indicate that about 300 bases of the probe are complementary to 0.05% of total pluteus mRNA. There are approximately 2 x 10^5 Spec 1 mRNAs per embryo, or about 500 per cell in the dorsal ectoderm.

At gastrula, cells of presumptive dorsal ectoderm are not distinguishable by morphological criteria. However, hybridization of the Spec 1 probe to sections of 40-hr gastrulae shows that these mRNAs are strikingly localized (Fig. 3). Measurements of grain densities show that the signal to noise is about 17 for the labeled

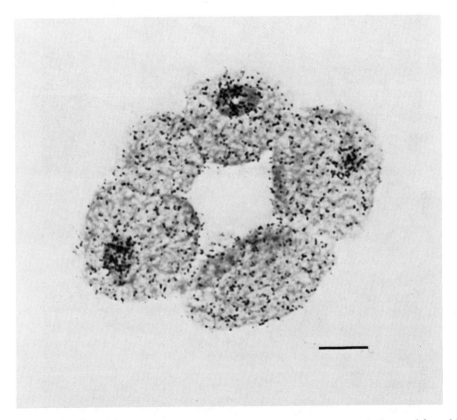

Figure 2: Reappearance of histone transcripts in nuclei at 16-cell
stage. Sections of 16-cell embryos were hybridized with the early
histone repeat RNA probe.

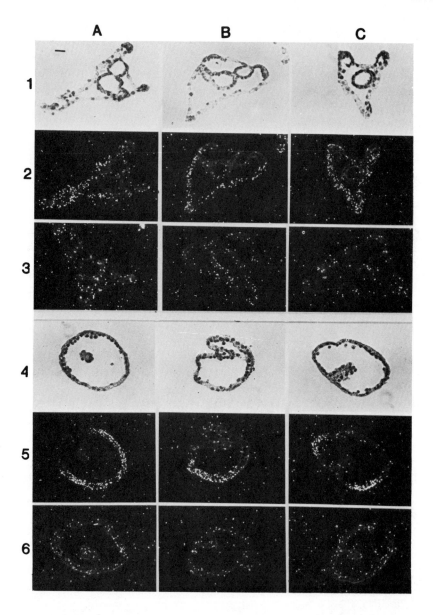

Figure 3: Localization of Spec 1 mRNAs in plutei and gastrulae. A probe complementary to Spec 1 mRNAs was hybridized to sections of plutei (rows 1 and 2) or gastrulae (rows 5 and 6). Sections in rows 1 and 4 are photographed under phase contrast; rows 2 and 5 show the same sections photographed under darkfield illumination. As a control for nonspecific binding, a nonhybridizing probe representing the mRNA strand was incubated under the same

area of ectoderm, while grain densities over "unlabeled" ectoderm,
archenteron, and primary mesenchyme are not different from levels of
background binding. Labeling appears to be essentially uniform over
presumptive dorsal ectoderm. In Fig. 4 is shown a tangential section
encompassing the border between labeled and unlabeled cells to
illustrate the sharpness of this margin. Cells of ectoderm either
express Spec 1 mRNAs at uniform concentration or have no transcripts
detectable by this assay. There is, for example, no evidence for a
dorsal-ventral gradient of Spec 1 mRNA content. Analysis of serial
sections allowed us to construct an approximate fate map of dorsal
ectoderm at gastrula based on the content of these cell-type specific
mRNAs (Fig. 5). Our measurements indicate that there are about the
same number of (presumptive) dorsal ectoderm cells, 350, at this
stage as in plutei, implying that there is little cell division in
this lineage between these two stages. Thus, the change in spatial
pattern of distribution of labeled cells between these two stages
largely reflects flattening of cells of the dorsal ectoderm and
consequent expansion of this sheet to encompass about 75% of the
embryo surface. The change in shape of dorsal ectoderm cells is a
major morphogenetic event initiated around the time that the Spec 1
mRNAs (and proteins) reach high concentrations in these cells, and it
is possible that the Spec 1 proteins have some function in this
process.

Preliminary analysis of 28-hr embryos indicates that Spec 1
mRNAs are also localized at this stage. Although we have not
determined the distribution in detail, individual sections show
patterns similar to those observed at gastrula. It seems likely that
these mRNAs are restricted to presumptive dorsal ectoderm from the
time their concentration begins to increase at about 20 hr.

The timing of appearance and localization of Spec 1 mRNAs bears
an interesting relationship to several other events in the embryo.
Studies in the laboratory of Dr. L. Cohen (personal communication)
have shown that after about 21 hr of development, cells of dorsal
ectoderm cannot be labeled by continuous incubation in ^{3}H thymidine
and therefore have withdrawn from the division cycle. We have
carried out preliminary hybridizations of the early variant histone
probe at low stringency which allows hybridization to homologous
early variant as well as to at least some late variant mRNAs. At
gastrula, we observe frequently a large region of embryo ectoderm
that contains little detectable histone mRNA. While a direct

Figure 3 (continued):

conditions (rows 3 and 6). The sections of gastrulae are cut
approximately in the three perpendicular planes of symmetry. a)
anterior; p) posterior; d) dorsal; v) ventral, an) animal pole; v)
vegetal pole. The bar represents 10 μ. (From Lynn et al., 1983.)

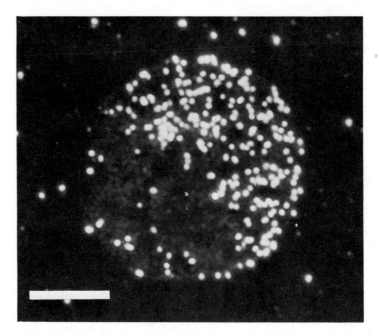

Figure 4: Margin of the presumptive dorsal ectoderm. A tangential
  section is shown which encompasses part of the border between the
  presumptive dorsal ectoderm and other regions of the ectoderm. The
  section was hybridized with the probe for Spec 1 mRNAs.
  Photographed under darkfield illumination.

correspondence has not yet been demonstrated, we believe that this
region of the embryo is the presumptive dorsal ectoderm. Although it
appears that these cells have largely terminated expression of early
variant genes and turned over their early variant mRNAs, further
studies are required to determine whether they are expressing any
late variant histone messages. Thermal denaturation of hybrids
formed in situ to RNAs in sections of plutei identifies them as late
variant messages, in agreement with other experiments showing that

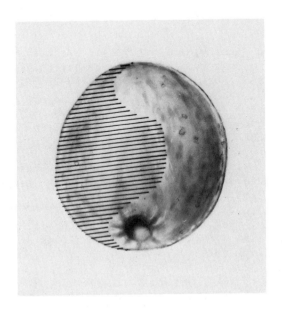

Figure 5: Fate map of the dorsal ectoderm in gastrulae. The shaded
area indicates the approximate region of presumptive dorsal
ectoderm as shown by content of Spec 1 mRNAs. an) animal;
ve) vegetal; d) dorsal; v) ventral. (From Lynn et al., 1983.)

histone mRNAs of plutei are predominantly late variant transcripts
(Newrock et al., 1978; Childs et al., 1979; Hieter et al., 1979).
Cells of dorsal ectoderm contain comparatively little histone mRNA,
while much higher concentrations are observed in growing arms, gut,
and, to a lesser extent, ventral ectoderm. Thus, the beginnings of
differentiation of dorsal ectoderm at about 20 hr of development
appear to involve withdrawal of cells from the division cycle,
cessation of synthesis, and degradation of early variant histone mRNA
and initiation of synthesis of Spec 1 mRNAs.

ACKNOWLEDGEMENTS

   This research was supported by grants from the National
Institutes of Health (GM25553) and by an Institutional Biomedical
Research Grant from USPHS.

REFERENCES

Angerer, L.M. and Angerer, R.C. (1981). Detection of poly(A)$^+$ RNA
   in sea urchin eggs and embryos by quantitative in situ
   hybridization. Nuc. Acids. Res. 9:2819-2840.

Brahic, M. and Haase, A.T. (1978). Detection of viral sequences of
   low reiteration frequency by in situ hybridization. Proc. Natl
   Acad. Sci. USA 75:6125-6129.

Brandhorst, B.P. (1976). Two-dimensional gel patterns of protein
   synthesis before and after fertilization of sea urchin eggs.
   Dev. Biol. 52:310-327.

Brandhorst, B.P. (1980). Simultaneous synthesis, translation, and
   storage of mRNA, including histone mRNA, in sea urchin eggs.
   Dev. Biol. 79:139-148.

Bruskin, A., Tyner, A.L., Wells, D.E., Showman, R.M., and Klein, W.H.
   (1981). Developmental regulation of six mRNAs enriched in
   ecoderm of sea urchin embryos. Dev. Biol. 87:308-318.

Butler, E. and Chamberlin, M.J. (1982). Bacteriophage SP6-RNA
   polymerase I. Isolation and characterization of the enzyme. J.
   Biol. Chem. 257:5772-5778.

Casey, J. and Davidson, N. (1977). Rates of formation and thermal
   stability of RNA:DNA and DNA:DNA duplexes at high concentrations
   of formamide. Nuc. Acids. Res. 4:1539-1552.

Childs, G., Maxson, R., and Kedes, L.H. (1979). Histone gene
   expression during sea urchin embryogenesis: isolation and
   characterization of early and late messenger RNAs of
   Strongylocentrotus purpuratus by gene-specific hybridization and
   template activity. Dev. Biol. 73:153-173.

Costantini, F.D., Scheller, R.H., Britten, R.J., and Davidson, E.H.
   (1980). Message sequences and short, repetitive sequences are
   interspersed in sea urchin egg poly(A)$^+$ RNAs. Nature
   287:111-117.

Crane, Jr., W.R., Durica, D.S., and Van Doren, K. (1981). Actin gene expression in developing sea urchin embryos. Mol. Cell. Biol. 1:711-720.

Davidson, E.H. (1976). Gene Activity in Early Development. Academic Press, New York.

Davidson, E.H., Hough-Evans, B.R., and Britten, R.J. (1982). Molecular biology of the sea urchin embryo. Sci. 217:17-26.

Diaz, M.O., Barsacchi-Pilone, G., Mahon, K.A., and Gall, J.G. (1981). Transcripts from both strands of a satellite DNA occur on lampbrush chromosome loops of the newt Notophthalmus. Cell 24:649-659.

Ernst, S.G., Hough-Evans, B.R., Britten, R.J., and Davidson, E.H. (1980). Limited complexity of the RNA in micromeres of 16-cell sea urchin embryos. Dev. Biol. 79:119-127.

Flytzanis, C.N., Brandhorst, B.P., Britten, R.J., and Davidson, E.H. (1982). Developmental patterns of cytoplasmic transcript prevalence in sea urchin embryos. Dev. Biol. 91:27-35.

Galau, G.A., Klein, W.H., Davis, M.M., Wold, B.J., Britten, R.J., and Davidson, E.H. (1976). Structural gene sets active in embryos and adult tissues of the sea urchin. Cell 7:487-505.

Hayashi, S., Gillam, I.C., Delaney, A.D., and Tener, G.M. (1978). Acetylation of chromosome squashes of Drosophila melanogaster decreases the background in autoradiographs from hybridization with ^{125}-labeled RNA. J. Histochem. Cytochem. 36:677-679.

Hentschel, C.C. and Birnstiel, M.L. (1981). The organization and expression of histone gene families. Cell 25:301-313.

Hieter, P.A., Hendricks, M.B., Hemminki, K., and Weinberg, E.S. (1979). Histone gene switch in the sea urchin embryo. Identification of late embryonic histone messenger ribonucleic acids and the control of their synthesis. Biochemistry 18:2707-2716.

Hough-Evans, B.R., Wold, B.J., Ernst, S.G., Britten, R.J., and Davidson, E.H. (1977). Appearance and persistence of maternal RNA sequences in sea urchin development. Dev. Biol. 60:258-277.

Hynes, R.O. and Gross, P.R. (1970). A method for separating cells from early sea urchin embryos. Dev. Biol. 21:383-402.

Kedes, L.H. (1979). Histone genes and histone messengers. Ann. Rev. Biochem. 48:837-870.

Lederman, L., Kawasaki, E.S., and Szabo, P. (1981). The rate of nucleic acid annealing to cytological preparations is increased in the presence of dextran sulfate. Anal. Biochem. 117:158-163.

Lynn, D.A., Angerer, L.M., Bruskin, A.M., Klein, W.H., and Angerer, R.C. (1983). Localization of a family of mRNAs to a single cell type and its precursors in developing sea urchin embryos. Proc. Natl. Acad. Sci. USA, in press.

Mauron, A., Levy, S., Childs, G., and Kedes, L.H. (1981). Monocistronic transcription is the physiological mechanism of sea urchin embryonic histone gene expression. Mol. Cell Biol. 1:661-671.

Mauron, A., Kedes, L.H., Hough-Evans, B.R., and Davidson, E.H. (1982). Accumulation of individual histone mRNAs during embryogenesis of the sea urchin Strongylocentrotus purpuratus. Dev. Biol. 94:425-434.

Maxson, R.E. and Wilt, F.H. (1981). The rate of histone mRNA synthesis during early sea urchin development. Dev. Biol. 83:380-386.

McClay, D.R. and Marchase, R.B. (1979). Separation of ectoderm and endoderm from sea urchin pluteus larvae and demonstration of germ layer-specific antigens. Dev. Biol. 71:289-296.

Newrock, K.M., Cohen, L.H., Hendricks, M.B., Donnelly, R.F., and Weinberg, E.S. (1978). Stage-specific mRNAs coding for subtypes of H2A and H2B histones in the sea urchin embryo. Cell 14:327-336.

Overton, C. and Weinberg, E.S. (1978). Length and sequence heterogeneity of the histone gene repeat unit of the sea urchin, S. purpuratus. Cell 14:247-258.

Rodgers, W.H. and Gross, P.R. (1978). Inhomogeneous distribution of egg RNA sequences in the early embryo. Cell 14:279-288.

Rogers, A.W. (1979). Techniques in Autoradiography. Elsevier/North Holland Biomedical Press, p. 143.

Scheller, R.H., McAllister, L.B., Crain, Jr., W.R., Durica, D.S., Posakony, J.W., Thomas, T.L., Britten, R.J., and Davidson, E.H. (1981). Organization and expression of multiple actin genes in the sea urchin. Mol. Cell Biol. 1:609-628.

Showman, R.M., Wells, D.E., Anstrom, J., Hursh, D.A., and Raff, R.A.
    (1982). Message-specific sequestration of maternal histone mRNA
    in the sea urchin egg. Proc. Natl. Acad. Sci. USA
    79:5944-5947.

Skoultchi, A. and Gross, P.R. (1973). Maternal histone messenger
    RNA: Detection by molecular hybridization. Proc. Natl. Acad.
    Sci. USA 70:2840-2844.

Tufaro, F. and Brandhorst, B.P. (1979). Similarity of proteins
    synthesized by isolated blastomeres of early sea urchin embryos.
    Dev. Biol. 72:390-397.

Venezky, D.L., Angerer, L.M., and Angerer, R.C. (1981).
    Accumulation of histone repeat transcripts in the sea urchin egg
    pronucleus. Cell 24:385-391.

Wells, D.E., Showman, R.M., Klein, W.H., and Raff, R.A. (1981).
    Delayed recruitment of maternal mRNA in sea urchin embryos.
    Nature 292:477-478.

Wetmur, J.G. and Davidson, N. (1968). Kinetics of renaturation of
    DNA. J. Mol. Biol. 31:349-379.

SUBCELLULAR LOCALIZATION OF MATERNAL HISTONE mRNAs AND

THE CONTROL OF HISTONE SYNTHESIS IN THE SEA URCHIN EMBRYO

Richard M. Showman, Dan E. Wells, John A. Anstrom,
Deborah A. Hursh, David S. Leaf, and Rudolf A. Raff[*]

ABSTRACT

Early α-subtype histone synthesis in the sea urchin embryo occurs using stored naturally synthesized mRNA. Embryonic α-subtype histone mRNAs do not appear until the 16-cell stage. We have examined these maternal α histone mRNAs and demonstrate that, unlike the bulk of maternal mRNAs, they are recruited into polysomes and translated after a 90-min to 2-hr delay following fertilization. Using cell fractionation procedures and in situ autoradiography, we further demonstrate that maternal α histone mRNAs are sequestered within the pronucleus of the mature egg. Appearance of α-subtype histone mRNAs in the polysomes of the zygote correlates with the dissociation of the nuclear envelope prior to first cleavage.

Maternal Messenger RNAs in Sea Urchin Eggs

Brachet et al. (1963), and Denny and Tyler (1964), demonstrated that sea urchin eggs that had been enucleated by centrifugation, and then artificially activated, carry out protein synthesis. Although those observations strongly suggested the existence of cytoplasmic mRNA in eggs, such activated physically enucleated eggs or merogones had only a limited ability to cleave and develop. At the same time Gross and Cousineau (1963, 1964), showed that functionally enucleated sea urchin embryos cultured in sufficient actinomycin D to block 95% of RNA synthesis, continued to replicate DNA, cleave, and undergo

[*]From the Program in Molecular, Cellular, and Developmental Biology, Department of Biology, Indiana University, Bloomington, IN 47405.

morphogenesis. These embryos finally suffered developmental arrest at the hatching blastula stage, but during the course of their development vigorously carried out protein synthesis (see also Greenhouse et al., 1971). Use of actinomycin to functionally enucleate sea urchin embryos has made possible the identification of several specific proteins which are the products of mRNAs synthesized during oogenesis but translated after fertilization, including tubulins (Raff et al., 1971, 1972), hyaline layer protein (Citkowitz, 1972), and histones (Ruderman and Gross, 1974). These stored mRNAs have come to be known collectively as maternal mRNAs.

In vitro translation studies of maternal mRNA began with the experiments of Slater and Spiegelman (1966). Using a cell-free system from E. coli to translate unfractionated RNA from sea urchin eggs, they compared the amount of synthesis directed by sea urchin RNA with that directed by viral RNA and estimated that 4% of unfertilized sea urchin egg RNA is mRNA. A similar experiment in which poly(A)$^+$ RNA was used to direct a cell-free system derived from sarcoma-180 cells was used by Jenkins et al. (1973), to arrive at a comparable estimate of 3% of sea urchin egg RNA as mRNA. While no specific products were isolated from these studies, the activity of egg mRNA in vitro suggested that at least a large fraction of the mRNA stored in the egg is perfectly competent for translation. Many other similar observations have strengthened this conclusion, including a number of studies on specific mRNAs, particularly those coding for the histones. Gross et al. (1973), have shown that RNA extracted from 20-40S mRNP particles of sea urchin eggs directs the synthesis of histones in vitro while Ruderman and Pardue (1977), in analyzing the classes of mRNA present in sea urchin eggs by cell-free translation, found poly(A)$^-$ histone mRNA, and poly(A)$^+$ and poly (A)$^-$ non-histone mRNAs. All these classes were translatable in vitro, with some qualitative differences seen between products of poly(A)$^+$ and poly(A)$^-$ non-histone mRNAs. In 1976 Lifton and Kedes showed that the non-translated, but competent histone mRNAs in sea urchin eggs are identical to histone mRNAs being translated after fertilization. They further showed that the electrophoretic mobility of histone mRNA isolated from unfertilized eggs is identical to that of well-characterized embryonic histone mRNAs. Furthermore, these histone mRNAs are translatable in vitro, producing well-defined histones characteristic of early development (Childs et al., 1979).

The now well defined program of synthesis and storage during oogenesis of maternal mRNAs, followed by their translation in early development, has given rise to the widely held hypothesis that maternal mRNAs provide the major part of the maternal program for early development, and further that their mRNAs govern synthesis of various cellular proteins and control early morphological events. This view has motivated much of the work done on maternal mRNAs. It is important to note that many potential modes of storage of information for a maternal developmental program exist, and that it has not yet been possible to demonstrate a role for maternal mRNA in controlling embryonic morphogenesis. Perhaps the strongest observations are those by Kalthoff (1983), and his co-workers on the anterior determinant of the diperan insect Smittia, which provide most of our current evidence that RNAs can encode morphogenetic information. No analagous data exist for sea urchin embryos, although morphological differentiation occurs quite early in development and includes the establishment of animal-vegetal polarity and determination of the size and fate of the blastomeres of the 16-cell stage embryo.

In studying the role of maternal mRNAs in developing embryos, it is necessary to consider both how the mRNA is stored and the manner in which it is utilized. Sea urchin embryos do not simply continue to translate the same set of mRNAs throughout development. Patterns of protein synthesis change, and most changes can be accounted for by the synthesis of new species of mRNA. Addition of new mRNAs by changes in RNA synthesis patterns is adequate at stages of development where sufficient nuclei are present to rapidly supply new mRNAs. If changes in protein synthesis patterns are required at early stages in development before many nuclei exist, other (post-transcriptional), controls must exist. One such alternate means of controlling the time of synthesis of particular proteins would be the selection of maternal mRNAs for translation in a sequence-specific manner. Such controls would be expected to operate during early transitions (maturation, fertilization, or onset of rapid cleavage), when the stored maternal mRNAs are heavily used and might involve proteins that must be synthesized rapidly and in large amounts. A control of this general type has been recently shown to operate in sea urchin embryos (Wells et al., 1981a,b; Showman et al., 1982). In this chapter we present recent results from our laboratory on the spatial organization and temporal regulation of translation of maternal histone mRNAs in the sea urchin embryo.

Timing of Recruitment of Maternal mRNAs into Embryo Polysomes

Figure 1 shows the kinetics of recruitment of egg ribosomes into embryo polysomes following fertilization. The formation of polysomes begins within a very few minutes of fertilization and reflects the recruitment of the mass of the maternal mRNAs of the egg. Raff et al. (1981) have constructed a quantitative computer model of the post-fertilization rise in total protein synthesis. Using this model, the kinetics of the rise can be accounted for by two major factors. The first is the rate at which maternal mRNA become available (by unmasking or other processes), for initiation. The second is the rate of ribosome transit of mRNAs. The initial recruitment process involves not only a large increase in the mass of mRNA found on the polysomes, but also a 50% decrease in the amount of those required for a given ribosome to complete the synthesis of a protein (Brandis and Raff, 1978; Hille and Albers, 1979). These two

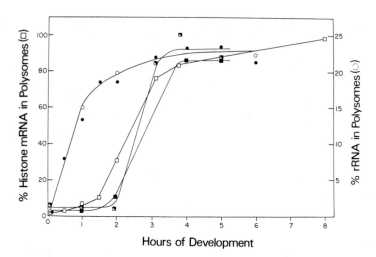

Figure 1: Time of appearance of rRNA and histone H1, H2B, and H3 mRNAs in the polysomes of S. purpuratus embryos at 13.5°C. (●)% rRNA in polysomes as determined by A_{253}; (○)% rRNA in polysomes as determined by hybridization with rDNA probe; % of histone H1 (■), H2B (▨), and H3 (▢), mRNA in polysomes as determined by cloned DNA probes (Wells et al., 1981 a, b).

factors appear to adequately account for the gross quantitative changes in protein synthesis that occur at fertilization. To address possible qualitative controls, other approaches are necessary.

The pattern of prevalent proteins synthesized both in the egg and up to the blastula stage, where newly synthesized mRNA begins to dominate the mRNA pool, appears to have, at best, only minor quantitative differences (Brandhorst, 1976). An exception to this general pattern has been found in the control of α subtype histone synthesis (Wells et al., 1981a and b). Using both Northern gel analysis of the amount of α histone mRNA that enters polysomes during development and short protein labeling periods combined with acrylamide gel electrophoresis that will resolve histones, we have been able to demonstrate that synthesis of the α-subtype histones does not begin at fertilization. Nevertheless, the mRNAs for these proteins are present in the pool of maternal mRNAs and can be readily detected by hybridization or by in vitro translation techniques (Wells et al., 1981a; Winkler and Steinhardt, 1981), (see Figs. 1 and 2). Titration of α-subtype maternal histone mRNA behavior with clones specific for H1, H2B, and H3 coding sequences reveals histone mRNAs are present in polysomes at only very low levels until 1.5 to 2 hr after fertilization. This time corresponds to about the period of the first cleavage at the temperature (13.5°C), used in these experiments. Between 2 and 3.5 hr after fertilization essentially all of the available histone α mRNA enters polysomes. Since no

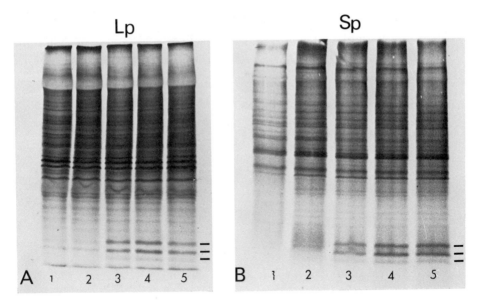

Figure 2: Appearance of ^{35}S-methionine in histones pulse labeled for
1-hr intervals between 1) 0-1 hr; 2) 1-2 hr; 3) 2-3 hr; 4) 3-4 hr;
and 5) 4-5 hr post-fertilization. Embryos were lysed directly in
SDS sample buffer according to Laemmli (1970), and the total
protein displayed on 10% acrylamide gels. Labeled proteins were
detected by autoradiography. Horizontal bars indicate positions of
marker histones (H3, H2A/H2B, H4).

synthesis of histone mRNAs is detectable prior to about 4-5 hr of
development (Wells et al., 1981a; and unpublished data), this mRNA
must be of maternal origin.

     The unique message-specific pattern of α histone mRNA
recruitment seen in Fig. 1 clearly demonstrates that instead of
exhibiting only a single, early mobilization of maternal mRNAs at
fertilization sea urchin embryos recruit maternal mRNAs in two
temporally distinct phases. The early event consists of the
mobilization of thousands of mRNA sequences, each present in moderate
to low prevalence; the second is limited to a few specific, highly
prevalent sequences. Although both events are set in motion by
fertilization, both the timing and sequence specific nature of the
second event indicate that quite distinct mechanisms are involved.

     Attempts to determine what factors might be associated with the
delayed expression of the α histone mRNA have met with little success
(see Table 1). Activation of Lytechinus pictus eggs with ammonia or
Strongylocentrotus purpuratus eggs with ammonia plus hypertonic sea
water shock has no effect on the timing of α histone mRNA appearance
in the polysomes. Similarly, although a slight delay in α histone

Table 1.   Percentage of Maternal α-Histone in Polysomes

		1 hr embryos	3-4 hr embryos
I.	Control Fertilized Embryos	6%	84%
II.	Artificially Activated Eggs		
	NH$_3$	ND	70%
	A23187	ND	85%
	Procaine	ND	70%
I·II.	Treated Embryos		
	Aphidicolin	ND	97%
	Actinomycin D	18%	88%
	Azetidine	ND	62%
	β-OH Leucine	ND	36%[*]
	Cytochalasin B	10%	45%[*]
	Colchicine	13%	88%
	Dimethyaminopurine	8%	40%[*]

ND--Not Determined
[*]General protein synthesis as reflected by rRNA recruitment into
   polysomes also depressed.

mRNA recruitment occurs upon activation of eggs with the calcium
ionophore A23187, a similar delay occurs in other processes as
indicated by delayed nuclear envelope breakdown. Procaine activation
likewise does not change recruitment timing. Attempts to disrupt
recruitment by altering the cytoarchitecture of fertilized eggs with
either 4 mM colchicine or 10 µg/ml cytochalasin B (concentrations
sufficient to disrupt both microtubules and microfilaments), fail to
either block recruitment or induce premature loading of the α histone
mRNAs, suggesting that these components do not play a major role in
the process. To try and determine if newly synthesized proteins
might control the delayed release, embryos of S. purpuratus were
cultured in the presence of either 75 µg/ml azetidine or 75 µg/ml
β-OH leucine, amino acid analogues known to produce defective
proteins. Although both compounds have broad non-specific effects on
general embryonic development (e.g., no cleavage occurs), movement of
α histone mRNAs into polysomes nonetheless occurs by 3 hr
post-fertilization.

A close correlation of histone mRNA synthesis and DNA replication has been noted by many investigators in several cell types. Both mammalian cell culture lines (Spalding et al., 1966), and yeast cells (Moll and Wintersberger, 1976), partly because of their ease of synchronization, have been extensively studied. In both cases, close correlation exists between DNA replication and histone synthesis. In the case of HeLa cells, restriction of histone synthesis to the S phase of the cell cycle appears to hold for certain histone subtypes and not for others. Wu and Bonner (1981), have shown that while the synthesis of major histones appears to be tied to DNA synthesis, a basal level of synthesis of other histone subtypes can be detected at all stages of the cell cycle.

In yeast, Hereford and co-workers (1981), have shown that transcription of both the H2A and H2B histone genes increases nearly 20 fold as the cell enters S phase. They have further been able, using strains of yeast that are defective for DNA replication at a non-permissive temperature, to demonstrate that blocking DNA synthesis induces a rapid, specific degradation of histone mRNA. Such a phenomena has also been reported by Borun et al. (1967, 1975), who noted a loss of histone mRNA from the polysomes of HeLa cells following the blocking of DNA replication. Thus it appears for some cell types that coordination of DNA replication and histone synthesis occurs and is achieved at both the transcriptional and post-transcriptional levels.

This close coupling is clearly not the case for some of the maternal histone mRNA in the sea urchin where continued translation of the message occurs independent of at least the first two cycles of DNA replication (Arceci and Gross, 1980). Since proteins were analyzed on one-dimensional acrylamide gels in their studies, Arceci and Gross were unable to distinguish between α histones and the cleavage stage (CS), histones, which are synthesized during the first three cell divisions. The persistence of histone synthesis during the first two cell cycles, a time when there is neither significant polysomal α histone mRNA (Wells et al., 1981a), nor α histone synthesis in activated anucleate merogones, suggests that the histone variants Arceci and Gross observed were predominantly cleavage stage histones. These are known to be synthesized in the unfertilized egg where no DNA replication is occurring (Herlands et al., 1982). Thus the possibility remained that the initial release of the maternal α mRNA might be tied to the DNA replication cycle.

To examine this we took 4-hr-old embryos in which DNA synthesis had been blocked by the DNA polymerase inhibitor aphidicolin (Brachet and DePetrocellis, 1981), and measured the percentage of the total α histone mRNA that is polysomal. At 4 hr post-fertilization, 97% of the α histone mRNA has been recruited into the polysomes of the treated embryos, a percentage consistent with the idea that the blockage of DNA replication (< 10% of controls as measured by ^{3}H-thymidine incorporation), has no effect on maternal α histone mRNA recruitment. Thus, while later cycling of the maternal α mRNA cannot be excluded, it is clear that neither initial recruitment nor half life of the stored α mRNA is affected when DNA replication is blocked. Similar uncouplings of DNA synthesis and histone gene expression have been described in the embryos of Xenopus (see Woodland, 1980, for review), and Drosophila (Anderson and Lengyel, 1980).

Localization of Maternal Histone mRNAs

The possibility that histone mRNAs might be localized within the egg was first suggested by the results of Venezky et al. (1981). In situ hybridization of cloned histone probes to sections of sea urchin eggs revealed a significant concentration of silver grains over the nucleus. After taking into account the small volume occupied by the nucleus and the low cytoplasmic signal, Venezky et al. (1981), calculated that approximately 12% of the total oocyte α histone mRNA was concentrated in the nucleus.

We (Showman et al., 1982) have tested for a nuclear association for these mRNAs in an alternate manner. When sea urchin eggs are subjected to a centrifugal force of 10,000 to 20,000 x g in a sucrose-sea water step gradient their contents first stratify, and then the eggs split into nucleate and anucleate fragments (Fig. 3). Both types of fragments can be separated and fertilized and will produce normal, albeit diminutive embryos. Nucleated halves can be further split to produce nucleate and enucleate quarters. Each fragment can be probed to determine the amount of a given RNA species each contains. The results of some of these experiments are shown in Table 2. All numbers in the table are set relative to the corresponding value for the whole egg. Total RNA (primarily ribosomes), and poly(A)$^+$ RNA (the bulk of maternal mRNAs in cytoplasmic mRNPs), distribute in approximate proportion to relative volumes. This is not unexpected since the centrifugal forces used would not be expected to displace particles of the size of ribosomes and mRNPs. A slightly lower than expected amount of RNA in the nucleated quarters is probably the result of the inclusion in the nucleated half and quarter of the large oil droplet that accumulates at the centripetal pole of the stratified egg. When the relative concentration of two moderately prevalent maternal mRNAs, actin and α-tubulin, were examined it was observed that they too distribute themselves in a manner expected of mRNAs present as free cytoplasmic mRNPs. Not all RNAs exhibit this general pattern, however. Mitochondrial rRNA (Wells et al., 1982), is displaced to the "heavy" enucleate half egg, in accord with the displacement of mitochondria.

Unlike the total RNA and mitochondrial rRNAs, the α-subtype maternal histone mRNAs distribute themselves in a unique manner. These mRNAs are found exclusively in nucleated fragments. Such a

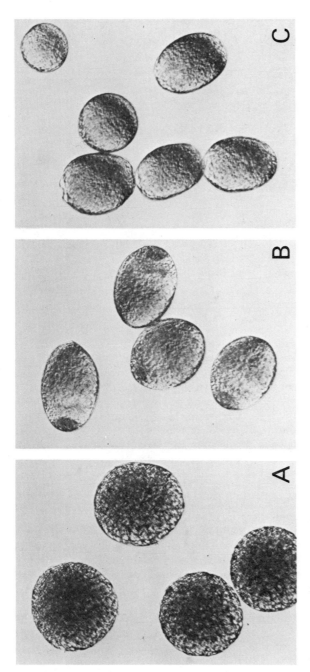

Figure 3: Centrifugally split S. purpuratus eggs. A) Whole, uncentrifugated eggs; B) stratified light halves containing the nucleus; C) heavy halves.

Table 2. Distribution of RNAs in Centrifugally Split Egg Fragments

	Whole Egg	Anucleate Half	Nucleate Half	Anucleate Quarter	Nucleate Quarter
Relative Volume	1.0	0.39	0.61	0.21	0.40
Total RNA	1.0	0.33	0.67	0.38	0.29
Poly(A)$^+$	1.0	0.31	0.69	0.39	0.30
Actin mRNA	1.0	0.33	0.67	(0.29)	(0.38)
α-tubulin mRNA	1.0	0.30	0.70	0.46	0.24
Histone mRNAs	1.0	0.02	0.98	< 0.01	0.98
Mitochondrial RNA	1.0	0.77	0.23	0.23	0.005

Egg and fragment volumes were calculated by using optical
measurements of the diameters of cells. Total RNA amounts were
determined by absorbance at 260 nm. Poly(A)$^+$ RNA values were
obtained by hybridization with ^{3}H-polyuridylate. Histone mRNA
probes used were H1, H2B, and H3 genes. Specific RNA measurements
were done using ^{32}P nick-translated sequence specific probes to
measure the amount of each RNA present. All values are expressed
as fraction of total per egg and are the average of two or more
separate experiments. The values in parentheses were determined
from a single experiment. Reproduced from Showman et al. (1982).

distribution is seen for at least three species of sea urchins (Table
3). Various treatments have been applied to eggs prior to splitting
to attempt to induce redistribution of the α histone mRNA (Table 3).
Neither disruption of microtubules or microfilaments with
cytochalasin B or colchicine nor elevated temperature had any effect.
Likewise, splitting of embryos 20 min post-fertilization left the α
histone mRNAs in the nucleate fragment. In contrast, fertilized eggs
which were cultured for 100 min until nuclear envelope breakdown had
begun and then split, displayed the appearance of α histone mRNA in
the anucleate fragments. The restriction of α histone mRNA to the
nucleate fragment implies the possibility of a nuclear localization
or the presence of organelles that contain histone mRNAs and are
displaced in a similar manner to nuclei by centrifugation. As a
third alternative, the histone mRNAs might be associated with a

Table 3. Distribution of α-Histone mRNA in Centrifugally Split
          Eggs and Embryos

		Nucleate Half	Anucleate Half
I.	Species (eggs)		
	Arbacia punctulata	99%	1%
	Strongylocentrotus droebachiensis	99%	1%
	Strongylocentrotus purpuratus	99%	1%
II.	Treated S. purpuratus eggs		
	Cytochalasin B--1 hr	99%	1%
	Colchicine--1 hr	99%	1%
	Elevated Temperature--15°C	99%	1%
III.	S. purpuratus Embryos		
	20 min post-fertilization (intact nucleus)	99%	1%
	100 min post-fertilization (no nucleus)	74%	26%
	3 hr post-fertilization + Dimethylaminopurine (intact nucleus)	74%	26%

particular region of the cortex of the egg. This last alternative
seems less likely for although animal-vegetal polarity of the sea
urchin embryo is already established in the unfertilized egg
(Horstadius, 1937; Schroeder, 1980), both the location of the nucleus
in the normal egg (Wilson, 1937), and the centrifugal/centripital
axis of centrifuged eggs (Morgan, 1927), seem to be random relative
to the rigid cortex. Thus it seems unlikely that the nucleus of
centrifuged eggs would be consistently displaced to an hypothetical
histone mRNA-bearing region of the cortex.

    To examine the two non-nuclear possibilities, we have split eggs
by an alternate method. Eggs were manually cut with a glass knife to
produce nucleate and enucleate halves (Fig. 4). In these eggs the
contents have not been systematically reorganized by the
centrifugation and splitting process. Furthermore, since cutting is
at random, a regular association of nuclei with any particular region
of cortex is eliminated. Any fragment thus produced should have an
equal probability of containing either the specific cortical region

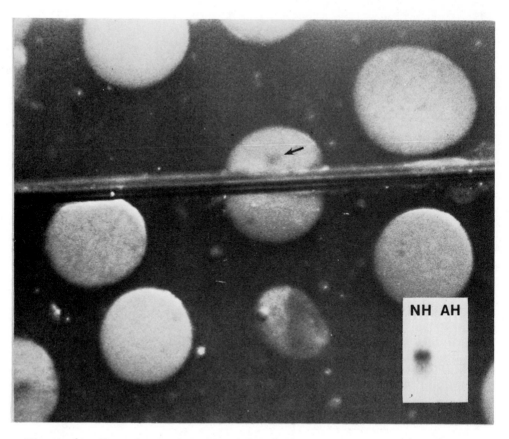

Figure 4: Manual cutting of S. droebachiensis egg. Each egg is cut
by hand and the nucleate and anucleate fragments collected in
separate depression slides. Arrow indicates the nucleus. Inset
shows an autoradiograph of total RNA from 35 nucleate (NH), and
anucleate (AH), fragments which were separated on agarose gels,
blotted to nitrocellulose, and hybridized with a ^{32}P-lobed H3 cDNA
probe.

or the theoretical histone mRNP containing organelles. Because of
the high prevalence of the histone mRNA in the sea urchin egg, as few
as ten eggs contain enough RNA to detect and quantitate by Northern
analysis (Fig. 4). When this is done, only the nucleated fragments
contained α histone mRNAs. In contrast, mitochondrial rRNA (Wells et
al., 1982), is equally distributed in both nucleate and anucleate
fragments. Thus it seems clear that the α histone mRNA is neither
located at some particular site in the cortex of the egg nor is it
restricted to some displacable organelle other than the nucleus.

It would also be expected that if histone mRNAs were contained in displacable cytoplasmic organelles, they should be observable by in situ hybridization to sections of centrifugally stratified eggs with a histone sequence probe. The results of such an experiment conducted jointly with Robert and Lynn Angerer at the University of Rochester are presented in Fig. 5. Although there is strong hybridization over the nucleus of stratified eggs, no significant concentration of grains over background is seen in the cytoplasm. Furthermore, the absence of a halo around the nucleus argues against a labile association of the mRNA with the outer surface of the nucleus. In combination with the other data discussed here, the uniform density of grains over the section of the nucleus makes it most probable that all of the maternal α-subtype histone mRNAs of the sea urchin egg are located in the female pronucleus and that the time of release of these mRNAs into the cytoplasm regulates the initiation of their translation in the embryo.

Although we have so far discussed the expression of only the α subtype histones, it should be noted that the maternal mRNAs for CS-subtype histones appear not to be sequestered into any displacable structure, but rather are distributed in the cytoplasm in the same manner as other non-histone mRNAs. By in vivo labeling we have been able to identify, via two-dimensional gel electrophoresis, the histones synthesized in each half. The CS-subtype histones are synthesized in roughly equal amounts by both nucleate and enucleate half eggs (J. Anstrom, unpublished data).

Release of Histone mRNAs from the Nucleus

As can be seen in Fig. 1, recruitment of maternal α-subtype histone mRNAs into polysomes begins about 30 min after nuclear breakdown preceeding the first mitotic cleavage of the zygote. Once the process starts, the rapid and nearly complete movement of histone mRNAs into polysomes within approximately 1 to 1.5 hr following the start of recruitment is observed. This rapid recruitment correlates well with the observations of Venezsky et al. (1981), who noted the absence of silver grains over nuclei of two-cell embryos examined by in situ hybridization. Such a correlation between nuclear membrane breakdown and the timing of recruitment of the α histone mRNA suggests a mechanistic connection.

To determine if nuclear membrane breakdown is required before α histone mRNA recruitment can occur, we have inhibited nuclear membrane breakdown with either dimethylaminopurine or aphidicolin (Showman et al., 1982). A concentration of 0.4 mM dimethylaminopurine reversibly inhibits both nuclear membrane breakdown and cleavage (Rebhun et al., 1973). Upon removal of the inhibitor, embryos treated with this compound will resume cleavage and ultimately produce apparently normal plutei. The effect of dimethylaminopurine on recruitment of α-subtype histone mRNAs is

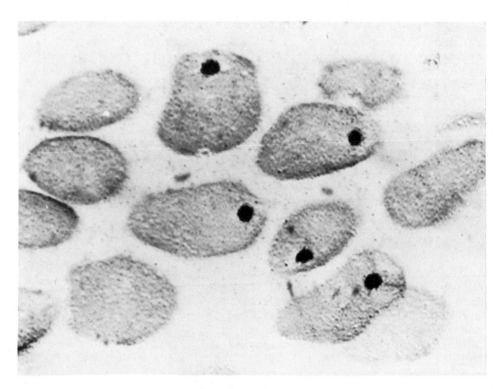

Figure 5:   In situ autoradiograph of stratified S. purpuratus eggs.
Eggs were prepared and autoradiographed as described by Venezky et
al. (1981).

shown in Fig. 6.   Despite the complete inhibition of nuclear membrane
breakdown, which normally occurs at 90 min (arrow in Fig. 6), the
onset of histone mRNAs occurs at the normal time.   Although
dimethylaminopurine does depress the rate of histone mRNAs
recruitment this probably reflects a non-specific effect on maternal
mRNA recruitment in general since polyribosome formation is also
depressed.   One of the side-effects of the DNA polymerase inhibitor
aphidicolin is the blocking of nuclear envelope breakdown (Brachet
and DePetrocellus, 1981).   As discussed earlier in relation to
blockage of DNA synthesis, the presence of aphidicolin has no effect
on the general pattern of α histone mRNA recruitment.   Thus it seems
possible that the release of maternal histone mRNAs from the nucleus
may occur by a process akin to the normal transport of mRNAs through
the nuclear membrane rather than by simple mechanical release.

Clearly the control of timing of release of α-subtype histone
mRNAs from the nucleus has not yet been resolved.   Nonetheless, the
specificity of mRNA release from the egg nucleus is a highly
significant point.   It has been clearly demonstrated (Ruderman and

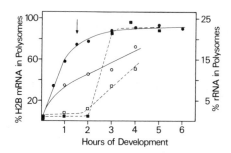

Figure 6: Recruitment of maternal histone H2B mRNA in the presence
   or absence of nuclear breakdown. Polysomes for appropriate time
   points were prepared from control cultures or cultures grown in the
   presence of 0.4 mM dimethylaminopurine. The percentage of total
   rRNA or histone H2B mRNA in the polysomal fraction was determined
   by absorbance at 254 nm or hybridization as previously described.
   Percentage of rRNA in polysomes as determined by absorbance at 254
   nm in control (●), and dimethylaminopurine (○), treated embryos.
   Percentage of H2B mRNA in polysomes as determined by titration with
   an H2B specific nick-translated probe in control (■), and
   dimethylaminopurine (□), treated embryos. Arrow indicates time
   of nuclear breakdown. Reproduced from Showman et al. (1982).

Schmidt, 1981), that unfertilized sea urchin eggs synthesize histone
mRNAs, and that these mRNAs appear in the polysomes of unfertilized
eggs. Herlands et al. (1982), have observed that although
unfertilized eggs contain both CS- and α-subtype histone mRNAs, only
those for CS-subtypes are translated in eggs. If it turns out that
the CS and α histone mRNAs are both synthesized coincidentally in the
egg, the localization of α-subtype histone mRNAs suggests that the
nuclei of unfertilized sea urchin eggs may have the ability to
selectively retain α-subtype histone mRNAs while simultaneously
releasing those for CS-subtypes to the cytoplasm.

Isolation of Nuclei and Histone mRNAs

The restriction of histone mRNAs to the nucleus as determined
using centrifugal splitting and in situ hybridization is inconsistent
with previously published data (e.g., Gross et al., 1973; Skoultchi
and Gross, 1973). These studies have consistently shown that the
bulk of the maternal histone mRNAs of the egg could be isolated in
cytoplasmic fractions. We (Showman et al., 1982) have isolated
nuclei under conditions that closely mimiced the osmotic and ionic
condition of the egg. Eggs were suspended in isotonic buffer and
disrupted by forcing them through a 10 μm Nitex mesh. This
homogenate was fractionated by successive centrifugations at
increasing g force. Pellets were collected at each centrifugation
step, nuclei were counted, and the RNA was extracted for

determination of histone mRNA content. Over 90% of the nuclei were
recovered in these pellets. Observation by light microscopy showed
these nuclei to be round, intact, and a uniform 10 µm in diameter.
The majority (78%) of the nuclei were recovered in the 150 x g
pellet. In contrast, 10% or less of α subtype histone mRNA remained
in this fraction; the bulk of α histone hybridizable material was
found in the 100,000 x g pellet. This g force would be expected to
pellet cytoplasmic mRNPs and, in fact, α tubulin mRNA, which is
characteristically found in a cytoplasmic mRNP particle exhibits a
uniform distribution in centrifugally split eggs (Table 2), was also
found primarily in this 100,000 x g pellet.

These data indicated that the association of the α subtype
histones with the nucleus was extremely labile to cell disruption,
even in the presence of buffers which ostensibly preserved nuclear
structure. Multiple variations in tonicity, salts, and speed of
isolation failed to change this basic result. We next attempted to
stabilize the α histone-nuclear association by isolating nuclei in
the presence of mild fixatives. Egg nuclei were subsequently
isolated in the presence of dilute ethanol according to the procedure
of Poccia et al. (1978). This isolation procedure allowed us to
isolate nuclei which retained their characteristic size and shape
(see Fig. 7). The number of nuclei or whole eggs was determined,
their RNA extracted, and the amount of α histone mRNA determined for
an equal number of nuclei or whole eggs. As Fig. 7 shows, all the α
histone mRNA found in approximately 5,000 eggs is contained in the
same number of fixed nuclei, demonstrating the localization of 100%
of the α histone in these ethanol fixed nuclei. Thus it would appear
that although α-subtype histone mRNAs are stably sequestered into the
nucleus in vivo, this association becomes extremely labile upon
disruption of the cell. Since the release of sequestered α-subtype
histone mRNAs in embryos occurs coincidently with nuclear envelope
breakdown at the first cleavage division, the simplest model of a
histone expression would argue for the release of the histone mRNA
from the nucleus as a direct result of its breakdown and not some
more sophisticated clock-dependent control system. The failure of
apparently intact nuclei to retain α-subtype histone mRNAs may, in
fact, represent an uncoupling of the mechanism(s) responsible for the
normal retention of mRNAs by the egg nucleus. We are currently
exploring other approaches for isolating nuclei that retain their
histone mRNA pools in order to examine such possibilities.

Possible Role of Histone mRNA Sequestration

The sequestration of α-subtype maternal histone mRNAs into the
nucleus and the subsequent delay in their release and translation in
the zygote is unique among mRNAs examined to date. One consequence
of this is a considerable functional constraint on the timing of
synthesis of α-subtype proteins. The presence of roughly 10^6 copies
of each α-subtype maternal mRNA and the extensive further production

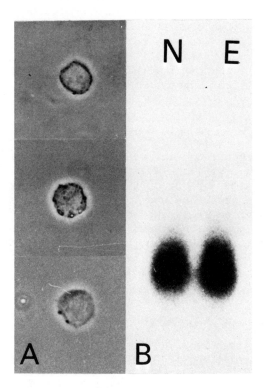

Figure 7: Restriction of histone mRNA to S. purpuratus egg nuclei
isolated following ethanol fixation. Nuclei were prepared using
the procedure of Poccia et al. (1978). Equal numbers of nuclei and
whole eggs (5,000), were phenol extracted, the RNA displayed on
agarose gels, blotted to nitrocellulose, and hybridized with
^{32}P-labeled DNA clone of the entire histone gene cluster. A)
isolated nuclei (1,000 X); B) northern blot of RNA from equal
numbers of nuclei (N), and whole eggs (E).

of α-subtype maternal mRNA, beginning at about 6 hr following
fertilization (Maxon and Wilt, 1982; Mauron et al., 1982), would seem
to argue for a major functional requirement in pre-blastula embryos
for this histone subtype. However, the discovery of significant
amounts of stored histone protein in the egg (Poccia et al., 1981;
Savic et al., 1981), and the development of small but normal plutei
from fertilized enucleate half eggs which lack the stored α histone
mRNA entirely, raise the question of the functional significance of
the maternal α histone mRNA pool. It is clear that after
fertilization the sperm chromatin undergoes a major remodeling and
incorporates specifically CS-subtype histones (Herlands et al.,
1982). Could sperm histones be replaced by α-variants? No
experimental answer is currently available, but the lengths to which
the egg goes to prevent premature synthesis of α-subtypes until

histone substitution is completed suggests the possibility that α
histones may interfere with the correct remodeling of male pronuclear
chromatin. The CS-subtypes might then serve in the transition to a
chromatin structure identical with that of the maternal chromatin
prior to entering first mitosis.

ACKNOWLEDGEMENTS

    We wish to acknowledge important technical assistance provided
by Carolyn Huffman. We thank William Klein for his advice and
collaboration in some aspects of this work and Robert and Lynne
Angerer for open discussions and for their collaboration in in situ
analysis of stratified eggs. We thank D. Stafford, L. Kedes, and E.
Weinberg for providing clones used in this work. Our work is
supported by grant HD-06902 from the National Institutes of Health.
DEW, DAH, and DSL were supported by a National Institutes of Health
Predoctoral Training Grant (GM 7227).

REFERENCES

Anderson, K.V. and Lengyel, J.A. (1980). Changing rates of
    histone synthesis and turnover in Drosophila embryos. Cell
    21:717-727.

Arceci, R.J. and Gross, P.R. (1980). Histone variants and
    chromatin structure during sea urchin development. Dev. Biol.
    80:186-209.

Borun, T.W., Scharff, M.D., and Robbins, E. (1967). Rapidly
    labeled polyribosome-associated RNA having the properties of
    histone message. Proc. Nat. Acad. Sci. USA 58:1977-1983.

Borun, T.W., Gabreilli, F., Ajiro, K., Zweidler, A., and Baglioni,
    C. (1975). Further evidence of transcriptional and
    translational control of histone messenger RNA during the HeLa
    S3 cycle. Cell 4:59-67.

Brachet, J., Deeroly, M., Ficq, A., and Quertier, J. (1963).
    Ribonucleic acid metabolism in unfertilized and fertilized sea
    urchin eggs. Biochim. Biophys. Acta 72:660-662.

Brachet, J. and DePetrocellis, B. (1981). The effects of
    adphidicolin, an inhibitor of DNA replication, on sea urchin
    development. Exp. Cell Res. 135:179-189.

Brandhorst, B.P. (1976). Two-dimensional gel patterns of protein
    synthesis before and after fertilization of sea urchin eggs.
    Dev. Biol. 52:310-317.

Brandis, J.W. and Raff, R.A. (1978). Translation of oogenetic mRNA in sea urchin eggs and early embryos. Demonstration of a change in translational efficiency following fertilization. Dev. Biol. 67:99-113.

Childs, G., Maxson, R., and Kedes, L.H. (1979). Histone gene expression during sea urchin embryogenesis: Isolation and characterization of early and late messenger RNAs of Strongylocentrotus purpuratus by gene-specific hybridization and template activity. Dev. Biol. 73:153-173.

Citkowitz, E. (1972). Analysis of the isolated hyaline layer of sea urchin embryos. Dev. Biol. 27:494-503.

Denny, P.C. and Tyler, A. (1964). Activation of protein biosynthesis in non-nucleate fragments of sea urchin eggs. Biochem. Biophys. Res. Commun. 14:245-249.

Greenhouse, G.A., Hynes, R.O., and Gross, P.R. (1971). Sea urchin embryos are permeable to actinomycin. Science 171:686-689.

Gross, P.R. and Cousineau, G.H. (1963). Effects of actinomycin-D on macromolecular synthesis and early development of sea urchin eggs. Biochem. Biophys. Res. Cummun. 10:321-326.

Gross, K.W., Jacobs-Lorena, M., Baglioni, C., and Gross, P.R. (1973). Cell-free translation of maternal messenger RNA from sea urchin eggs. Proc. Nat. Acad. Sci. USA 70:2614-2618.

Gross, P.R. and Cousineau, G.H. (1964). Macromolecule synthesis and the influence of actinomycin on early development. Exp. Cell Res. 33:368-395.

Hereford, L.M., Osley, M.A., Ludwig, J.R., and McLaughlin, C.S. (1981). Cell cycle regulation of yeast histone mRNA. Cell 24:367-376.

Herlands, L., Allfrey, V.G., and Poccia, D. (1982). Translational regulation of histone synthesis in the sea urchin Strongylocentrotus purpuratus. J. Cell Biol. 94:219-223.

Hille, M.B. and Albers, A.A. (1979). Efficiency of protein synthesis after fertilization of sea urchin eggs. Nature (London) 278:469-471.

Horstadius, S. (1937). Investigations as to the localization of the micromere-, the skeleton-, and the endoderm-forming material in the unfertilized egg of Arbacia punctulata. Biol. Bull. 73:295-316.

Jenkins, N.A., Taylor, M.W., and Raff, R.A. (1973). In vitro translation of oogenetic messenger RNA of sea urchins and picornavirus with a cell-free system from sarcoma-180. Proc. Nat. Acad. Sci. USA 70:3287-3291.

Kalthoff, K. (1983). Cytoplasmic determinants in dipteran eggs. In "Time, Space, and Pattern in Embryonic Development," W.R. Jeffery and R.A. Raff (eds.). A.R. Liss, Inc., N.Y.

Laemmli, U.K. (1970). Cleavage of structural proteins during the assembly of the head of bacteriophage T4. Nature (London) 227:680-685.

Lifton, R.P. and Kedes, L.H. (1976). Size and sequence homology of masked maternal and embryonic histone messenger RNAs. Dev. Biol. 48:47-55.

Mauron, A., Kedes, L.H., Hough-Evans, B., and Davidson, E.H. (1982). Accumulation of individual histone mRNAs during embryogenesis of the sea urchin Strongylocentrotus purpuratus. Dev. Biol., in press.

Maxson Jr., R.E. and Wilt, F.H. (1982). Accumulation of the early histone messenger RNAs during the development of S. purpuratus. Dev. Biol., in press.

Moll, R. and Wintersberger, E. (1976). Synthesis of yeast histones in the cell cycle. Proc. Nat. Acad. Sci. USA 73:1863-1967.

Morgan, T.H. (1927). "Experimental Embryology." New York: Columbia University Press.

Poccia, D.L., Levine, D., and Wang, J.C. (1978). Activity of a DNA topoisomerase (nicking-closing enzyme) during sea urchin development and the cell cycle. Dev. Biol. 64:273-283.

Poccia, D., Salik, J., and Krystal, G. (1981). Transitions in histone variants of the male pronucleus following fertilization and evidence for a maternal store of cleavage-stage histones in the sea urchin egg. Dev. Biol. 82:287-296.

Raff, R.A., Brandis, J.W., Huffman, C.J., Koch, A.L., and Leister, D.E. (1981). Protein synthesis as an early response to fertilization of the sea urchin egg: A model. Dev. Biol. 86:265-271.

Raff, R.A., Colot, H.V., Selvig, S.E., and Gross, P.R. (1972). Oogenetic origin of messenger RNA for embryonic synthesis of microtubule proteins. Nature (London) 235:211-214.

Raff, R.A., Greenhouse, G., Gross, K.W., and Gross, P.R. (1971). Synthesis and storage of microtubule proteins by sea urchin embryos. J. Cell Biol. 50:516-527.

Rebhun, L.I., White, D., Sander, G., and Ivy, N. (1973). Cleavage inhibition in marine eggs by puromycin and 6-dimethylaminopurine. Exp. Cell Res. 77:312-318.

Ruderman, J.V. and Gross, P.R. (1974). Histones and histone synthesis in sea urchin development. Dev. Biol. 36:286-298.

Ruderman, J.V. and Pardue, M.L. (1977). Cell-free translation analysis of messenger RNA in echinoderm and amphibian development. Dev. Biol. 60:48-68.

Ruderman, J.V. and Schmidt, M.R. (1981). RNA transcription and translation in sea urchin oocytes and eggs. Dev. Biol. 81:220-228.

Savic, A., Richman, P., Williamson, P., and Poccia, D. (1981). Alterations in chromatin structure during early sea urchin embryogenesis. Proc. Nat. Acad. Sci. USA 78:3706-3710.

Schroeder, T.E. (1980). Expressions of the prefertilization polar axis in sea urchin eggs. Dev. Biol. 79:428-443.

Skoultchi, A. and Gross, P.R. (1973). Maternal histone messenger RNA: Detection by molecular hybridization. Proc. Nat. Acad. Sci. USA 70:2840-2844.

Showman, R.M., Wells, D.E., Anstrom, J., Hursh, D.A., and Raff, R.A. (1982). Message-specific sequestration of maternal histone mRNA in the sea urchin egg. Proc. Nat. Acad. Sci. USA 79:5944-5947.

Slater, D.W. and Spiegelman, S. (1966). An estimation of the genetic messages in the unfertilized echinoid egg. Proc. Nat. Acad. Sci. USA 56:164-170.

Spalding, J., Kaijawara, K., and Mueller, G.C. (1966). An extracted basic protein isolated from HeLa nuclei and resolved by eleetrophoresis. Proc. Nat. Acad. Sci. USA 56:1535-1542.

Venezky, D.L., Angerer, L.M., and Angerer, R.C. (1981).
   Accumulation of histone repeat transcripts in the sea urchin
   egg pronucleus. Cell 24:385-391.

Wells, D.E., Bruskin, A.M., O'Brachta, D.A., and Raff, R.A.
   (1982). Prevalent RNA sequences of mitochondrial origin in
   sea urchin embryos. Dev. Biol. 92:557-562.

Wells, D.E., Showman, R.M., Klein, W.H., and Raff, R.A. (1981a).
   Delayed recruitment of maternal mRNA in sea urchin embryos.
   Nature 292:477-478.

Wells, D.E., Showman, R.M., Klein, W.H., and Raff, R.A. (1981b).
   Translational regulation in sea urchin embryos. Biol. Bull.
   161:322.

Wilson, E.B. (1937). "The Cell in Development and Heredity." New
   York: MacMillan Pub. Co., 3rd Edition.

Winkler, M.M. and Steinhardt, R.A. (1981). Activation of protein
   synthesis in a sea urchin cell-free system. Dev. Biol.
   84:432-439.

Woodland, H.R. (1980). Histone synthesis during the development
   of Xenopus. FEBS Ltrs. 121:1-7.

Wu, R.S. and Bonner, W.M. (1981). Separation of basal histone
   synthesis from S-phase histone synthesis in dividing cells.
   Cell 27:321-330.

A FAMILY OF mRNAs EXPRESSED IN THE DORSAL

ECTODERM OF SEA URCHIN EMBRYOS

William H. Klein, Lisa M. Spain, Angela L. Tyner,
John Anstrom, Richard M. Showman, Clifford D. Carpenter,
Elizabeth D. Eldon, and Arthur M. Bruskin[*]

INTRODUCTION

Several rationales are currently used for analyzing genes of potential importance during early development. The classical approach is mutational analysis. A wide variety of mutations in Drosophila (e.g., Nusslein-Vollard, 1979), C. elegans (e.g., Wood et al., 1980), amphibians (e.g., Malacinski and Brothers, 1974), mice (e.g., Bennett, 1980), and other organisms have profound affects on the developing embryo. Unfortunately in most cases, determining the actual function of the gene product in question is a difficult problem. Nevertheless, molecular studies of a few mutations, particularly the homoeiotic loci of Drosophila, have already advanced to the point where the genes and transcripts have been isolated and characterized (Marx, 1981).

A second approach to studying developmentally important genes is that of using conserved gene families which are already well characterized. Analyses using both genetic and molecular approaches have demonstrated the existence of stage and tissue specific subtypes within many gene families, including the histone (Newrock et al., 1978), actin (McKeown and Firtel, 1981), and tubulin (Kemphues et al., 1979) gene families. It is not clear why very closely related proteins are differentially expressed. It may be that each particular protein in a family has a very specific function, or alternatively that the differential expression is

---

[*]From the Program in Molecular, Cellular, and Developmental Biology, Department of Biology, Indiana University, Bloomington, IN 47405.

mainly the result of the organization of the family in the genome (Davidson, 1982).

Many laboratories, including our own, have used a third approach to early development: that of finding genes expressed at a specific time and place in the embryo but whose functions are unknown. The reasonable but necessary assumption of such an approach is that genes which display strict developmental regulation are likely to play important roles in embryogenesis. This approach has the advantage of not being limited to a specific mutational screen or to a protein whose basic function is already known. In theory any developmentally regulated gene can be isolated provided the sensitivity of the technique is high enough. The obvious disadvantage of the approach is that it may be very difficult to determine the function of the gene in question. It is also possible that the function of the gene, once determined, will have no developmental significance to the organism. The purpose of this article is to demonstrate the usefulness of studying developmentally regulated genes which have been previously uncharacterized.

We have used sea urchin embryos to isolate a family of ectoderm specific mRNAs (termed Spec mRNAs for Strongylocentrotus purpuratus ectoderm mRNAs) which accumulate in the late stages of sea urchin embryogenesis (Bruskin et al., 1981, 1982). In this article we summarize some of the features of these mRNAs, their genes, and their proteins. We present evidence that the mRNAs can be divided into two subfamilies and that the proteins encoded by at least one of the subfamilies are localized to a specific ectodermal cell type in the pluteus.

pSpec1 and pSpec2 cDNA Clones

Large numbers of sea urchin plutei can be easily fractionated into each of their primary germ layers, endoderm, mesoderm, and ectoderm (McClay and Chambers, 1978). We used this procedure and a library of pluteus cDNA clones to isolate tissue specific sequences from S. purpuratus plutei (Bruskin et al., 1981). Two of the cDNA clones we isolated, pSpec1 and pSpec2, code for two closely related mRNAs present in the pluteus ectoderm. The cloned sea urchin sequences are approximately 80% homologous and represent the 3' untranslated region of a 1.5 kb mRNA (Spec1) and a 2.2 kb mRNA (Spec2).

Accumulation During Embryogenesis of Spec1 and Spec2 mRNAs

We have previously reported that Spec1 and Spec2 are developmentally regulated mRNAs (Bruskin et al., 1982). The presence of both RNAs was monitored during embryogenesis by RNA gel blot analysis. The mRNAs are present in low or undetectable levels in the unfertilized egg and early cleaving embryo. By hatching blastula stage (20 hours) the Spec1 mRNAs begin to accumulate. Their levels increase over 100 fold from early cleavage to gastrula. The Spec2 mRNAs also accumulate during embryogenesis starting about ten hours later than the Spec1 mRNAs. Quantitation of the amount of Spec1 mRNAs by solution hybridization suggests they are about 0.6% of the embryo mRNA at 50 hours (Lynn et al., 1983). The Spec2 mRNAs are about 1/10 as prevalent at their maximum level, also at 50 hr (Bruskin et al., 1981). Thus while they are two very closely related mRNAs, Spec1 and Spec2 show distinct patterns of accumulation both in terms of time of appearance and absolute quantity.

Spec1 and Spec2 mRNA Families

We were able to show that the pSpec1 and pSpec2 cDNA clones select a family of related mRNAs by using the technique of hybrid-selected translation (Bruskin et al., 1982). RNA selected by either clone codes for approximately ten polypeptides with molecular weights ranging from 14,000 to 17,000 daltons and isoelectric points ranging from pH 5.0 to pH 6.0. Experiments comparing the in vitro translation products of Spec1 and Spec2 mRNAs with ectoderm proteins synthesized in vivo show that at least seven of the polypeptides comigrate on two-dimensional polyacrylamide gels. We concluded that Spec1 and Spec2 code for mRNAs which direct the synthesis of a family of similar but distinct ectoderm proteins.

To determine whether the Spec1 or Spec2 mRNAs each code for all of the low molecular weight polypeptides or whether each represents a distinct subset, we took advantage of the size difference between the two mRNAs. RNA from gastrula staged embryos was electrophoresed on methylmercury agarose gels and the gel was then fractionated into several pieces. The RNA was eluted from each gel piece and divided into two aliquots. One aliquot was re-electrophoresed, blotted to nitrocellulose filter paper, and probed with nick-translated pSpec2. The other aliquot was translated in a rabbit reticulocyte lysate in the presence of ^{35}S methionine and the radiolabeled proteins were displayed on two-dimensional polyacrylamide gels (Fig. 1). With unfractionated RNA (Fig. 1A, lane 1) the pSpec2 probe hybridizes strongly to transcripts 2.2 kb in length. The weaker bands represent hybridization to the related 1.5 kb Spec1 mRNAs and also to uncharacterized 3.2 kb transcripts presumably related in sequence to Spec1 and Spec2. Lanes 2 and 3 show two of the size fractionated RNAs, one containing the 2.2 kb and 3.2 kb transcripts and the other the 1.5 kb transcripts. The translation products corresponding to lanes 1, 3, and 2 from these fractions are displayed on the two-dimensional gels shown in Fig. 1B, 1C, and 1D, respectively.

The ectoderm proteins previously identified as belonging to the Spec family are numbered in the translation of unfractionated RNA depicted in Fig. 1B (Bruskin et al., 1982). It is evident from a comparison of the gels in Fig. 1B, 1C, and 1D that the 2.2 kb mRNAs

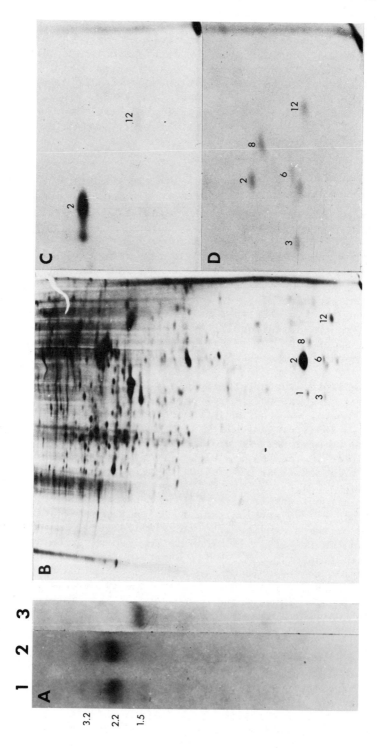

Figure 1: Fractionation of the 1.5 kb Spec1 mRNAs from the 2.2 kb
Spec2 mRNAs. Total RNA from gastrula staged embryos was
electrophoresed on 1% agarose methylmercury gels. The region of
the gel corresponding to RNAs 1 to 4 kb in length was divided into
five fractions and RNA from each fraction was isolated by
electroelution followed by ethanol precipitation. A) An aliquot of
the fractionated RNA was electrophoresed on a 1% agarose
formaldehyde gel, blotted to nitrocellulose filter paper, and
probed with ^{32}P Spec2 DNA labeled by nick translation: lane 1,
unfractionated RNA; lane 2, fraction 1 RNA; lane 3, fraction 5 RNA.
The numbers on the left indicate the size of the transcripts in kb.
B) Translation of unfractionated RNA in a rabbit reticulocyte
lysate. The ^{35}S labeled proteins were electrophoresed according to
O'Farrell (1975) and the resultant gel was fluorographed. The
numbers indicate the ectoderm proteins as defined in Bruskin et al.
(1982). C) Translation of fraction 1 RNA. Only the region of the
gel corresponding to the migration of the ectoderm proteins is
shown. D) Translation of fraction 5 RNA.

translate a different subset of ectoderm proteins than the 1.5 kb
mRNAs. The 2.2 kb mRNAs translate the five or six less abundant
polypeptides, while the 1.5 kb mRNAs appear to translate only the
most abundant protein. We conclude from these experiments that the
Spec1 and Spec2 mRNAs belong to two separate subclasses. The more
abundant Spec1 mRNAs appear to code for only a single protein
detectable among the in vitro translation products displayed on
two-dimensional gels. The Spec2 mRNAs, about 1/10 as prevalent as
the Spec1 mRNAs, code for several proteins. These proteins are of
similar size and charge to each other and to the Spec1 translation
product.

Localization of the Ectoderm Protein Coded by Spec1 mRNA

We have demonstrated previously that the Spec1 mRNAs are
localized to a specific ectodermal cell type in the pluteus stage
embryo (Lynn et al., 1983). In these experiments embryos were fixed,
sectioned, and hybridized in situ with a ^{3}H labeled probe
complimentary to Spec1 mRNA. The autoradiographs from these
experiments show hybridization to only the dorsal ectoderm cells of
the pluteus. These cells have a distinct morphology from the rest of
the pluteus ectoderm. They are squamous epithelial cells.
Hybridization to gastrula stage embryos shows grains localized
strictly to a specific set of cells that are morphologically
indistinguishable from the other ectodermal cells in the gastrula
(Lynn et al., 1983). Thus the Spec1 mRNAs accumulate in cells
determined to become dorsal ectoderm cells. This cell type
apparently ceases cell division sometime during the late

blastula-early gastrula stage of development (L. Cohen, personal communication) and therefore appears to represent a terminally differentiated cell type.

To determine where the ectoderm proteins are located in the embryo, antibodies against the most abundant ectoderm proteins (spots 1 and 2 of Fig. 1B) were produced. The antigen was prepared by electrophoresing pluteus stage proteins on two-dimensional gels, staining the gels with Coomassie Blue, and eluting the appropriate stained spots from the gel. Ectoderm proteins 1 and 2 were purified in this manner, pooled, and injected into a three-month-old Balb/C female mouse.

The antiserum isolated from the mouse had a high titer of antibody against ectoderm proteins 1 and 2 as judged by a Western protein blot analysis (Tyner and Klein, unpublished data). In the Western analysis a two-dimensional polyacrylamide gel containing pluteus protein was blotted to nitrocellulose filter paper. The nitrocellulose paper was reacted first with antiserum and then with a peroxidase conjugated antibody. Only proteins 1 and 2 were bound by the primary antiserum.

To localize the ectoderm proteins in the pluteus, embryos were fixed in ethanol, embedded in paraffin, sectioned, and reacted with the antiserum. Following this reaction the embryos were incubated with the secondary antibody followed by diaminobenzedine and hydrogen peroxide (Fig. 2). The reaction product is localized to the dorsal ectoderm (Fig. 2A). There is no precipitate in any of the gut tissue or in the ventral part of the animal (i.e., the mouth and oral and anal arms). The pre-immune serum shows no precipitate in any cells of the embryo (Fig. 2B). It appears that the ectoderm proteins are within the dorsal ectoderm cells. They do not seem to be extracellular.

It is clear from Fig. 2 that the ectoderm proteins are localized to the same cell type as the Spec1 mRNAs. Earlier analysis showed that the ectoderm proteins accumulate substantially after gastrulation (Bruskin et al., 1982). There is, therefore, a strong correlation between the time of morphological differentiation of these cells and the maximum accumulation of the ectoderm proteins.

In our initial attempts to purify some of the ectoderm proteins from pluteus stage embryos, we have lysed the embryos in non-ionic detergent (Tyner and Klein, unpublished data). Following centrifugation the most abundant ectoderm proteins are greatly enriched in a 5000 xg pellet. This result is also obtained starting with isolated pluteus ectoderm cells. An intriguing possibility is that the proteins are either in the ectoderm nucleus or that they are part of some cytoskeletal structure. By using the antiserum coupled

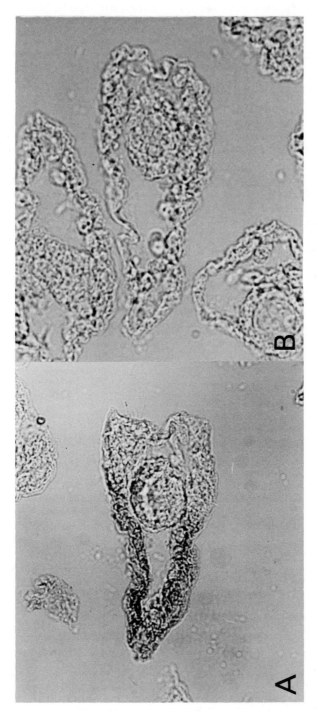

Figure 2: Immunostaining of S. purpuratus pluteus sections with antiserum against the major ectoderm proteins. The mouse antibodies were visualized by reacting the tissue with goat anti-mouse antibodies conjugated with peroxidase. The peroxidase activity was dectected by addition of diaminobenzidine and hydrogen peroxide to the tissue sections. A) Antiserum. B) Pre-immune serum.

with transmission electron microscopy, we should now be able to
distinguish between these two possibilities.

Structure of the Spec mRNAs and Their Genes

Our preliminary results indicate that the Spec1 1.5 kb mRNAs are
probably a small group of six or related variants (Bruskin,
Carpenter, and Klein, unpublished data). These variants are encoded
by six or seven different Spec1 genes. The Spec1 mRNAs are unusual
in their structure. They can be divided into three separate regions
corresponding to at least three exons in the Spec1 genes. The 3' end
corresponds to exon 1 and is 840 bases in length. Exon 1 of at least
one of the genes contains no translational reading frame. Exon 1 of
the Spec1 genes examined to date is extremely conserved (> 95%
homology). This conservation is unusual for the 3' untranslated
regions of eukaryotic mRNAs. Moreover, exon 1 contains a highly
repeated element about 200 to 300 bases from its 3' end. This
repeated element is 150 bases in length and is present at 2000 to
3000 other sites in the sea urchin genome. The elements present in
two of these other sites have been sequenced and are each about 80%
homologous to the elements present in the Spec1 genes. The large
majority of these other genomic sites do not code for Spec genes.
The function, if any, of the repetitive element is unknown, although
it seems likely it represents an insertion event into the ancestral
Spec genes before the genes themselves multiplied (Carpenter et al.,
1982).

The second region of the Spec1 mRNAs is 110 to 120 bases long
and corresponds to exon 2. We have sequenced several exon 2 from
several clones and our preliminary results indicate there is less
homology (90%) among exon 2 sequences than exon 1 sequences.

The final region of the Spec1 mRNAs contains the remaining 600
bases. This corresponds to at least one more exon in the Spec1
genes, although it is not clear if it is split into several more.
This region must contain the 5' end of the mRNAs and most of the 450
bases of coding sequence necessary for the 17,000 dalton ectoderm
protein.

We have not yet analyzed the Spec2 genes, but from the analysis
shown in Fig. 1 it is clear that there must be several distinct Spec2
mRNAs. It seems reasonable to assume that these RNAs must be
trancribed from several different genes.

ACKNOWLEDGEMENTS

This study was supported by an NIH grant, PHS HD14182, to W.H.K.
A.M.B. and C.D.C. are supported by a predoctoral training grant from
the NIH.

REFERENCES

Bennett, D. (1980). The T-complex in the mouse: an assessment after 50 years of study. The Harvey Lectures, Series 74:1-21.

Bruskin, A.M., Bedard, P.A., Tyner, A.L., Showman, R.M., Brandhorst, B.P., and Klein, W.H. (1982). A family of proteins accumulating in ectoderm of sea urchin embryos specified by two related cDNA clones. Develop. Biol. 91:317-324.

Bruskin, A.M., Tyner, A.L., Wells, D.E., Showman, R.M., and Klein, W.H. (1981). Accumulation in embryogenesis of five mRNAs enriched in the ectoderm of the sea urchin pluteus. Develop. Biol. 87:308-318.

Carpenter, C.D., Bruskin, A.M., Spain, L.M., Eldon, E.D., and Klein, W.H. (1982). The 3' untranslated regions of two related mRNAs contain an element highly repeated in the sea urchin genome. Nucleic Acids Res. 10:7829-7842.

Davidson, E.H. (1982). Evolutionary change in genomic regulatory organization: Speculations on the origins of novel biological structure. In "Evolution and Development," J.T. Bonner (ed.), pp. 65-84. Dahlem Konferenzen, Berlin, Heidelberg, New York: Springer-Verlag.

Kemphues, K.J., Raff, R.A., Kaufman, T.C., and Raff, E.C. (1979). Mutation in a structural gene for a β-tubulin specific to testis in Drosophila melanogaster. Proc. Natl. Acad. Sci. USA 76:3991-3995.

Lynn, D.A., Angerer, L.M., Bruskin, A.M., Klein, W.H., and Angerer, R.C. (1983). Localization of a family of mRNAs in a single cell type and its precursors in sea urchin embryos. Proc. Natl. Acad. Sci. USA (in press).

Malacinski, G.M. and Brothers, A.J. (1974). Mutant genes in the Mexican axolotl. Science 184:1142-1147.

Marx, J.L. (1981). Genes that control development. Science 213:1485-1488.

McClay, D.R. and Chambers, A.F. (1978). Identification of four classes of all surface antigens appearing at gastrulation in sea urchin embryos. Develop. Biol. 63:179-186.

McKeown, M. and Firtel, R.A. (1981). Evidence for subfamilies of actin genes in Dictyostelium as determined by comparison of 3' end sequences. J. Mol. Biol. 151:593-604.

Newrock, K.M., Cohen, L.H., Hendricks, M.B., Donnelly, R.J., and
    Weinberg, E.S. (1978). Stage specific mRNAs coding for
    subtypes of H2 and H2B histones in the sea urchin embryo. Cell
    14:327-336.

Nusslein-Volhard, C. (1979). Maternal effect mutations that alter
    the spatial coordinates of the embryo of Drosophila
    melanogaster. In "Determinants of Spatial Organization," S.
    Subtenlny and I. Konigsberg (eds.), pp. 185-211. Academic
    Press, New York.

O'Farrel, P.H. (1975). High resolution two-dimensional
    electrophoresis of proteins. J. Biol. Chem. 250:4007-4021.

Wood, W.B., Hecht, R., Carr, S., Vanderslice, R., Wolf, N., and
    Hirsh, D. (1980). Parental effects and phenotypic
    characterizations of mutations that affect early development in
    Caenorhabditis elegans. Develop. Biol. 74:469.

SEGREGATION OF GERM-LINE-SPECIFIC ANTIGENS

DURING EMBRYOGENESIS IN CAENORHABDITIS ELEGANS

Susan Strome and William B. Wood[*]

ABSTRACT

   Germ cells in a wide variety of invertebrate and vertebrate
species contain distinctive cytoplasmic organelles that have been
visualized by electron microscopy. The ubiliquity of such structures
suggests that they play some role in germ-line determination or
differentiation, or both. However, the nature and function of these
structures remain unknown. We describe experiments with two types of
immunologic probes, rabbit sera and mouse monoclonal antibodies,
directed against cytoplasmic granules that are unique to germ-line
cells in the nematode, Caenorhabditis elegans, and that may
correspond to the germ-line-specific structures seen by electron
microscopy in C. elegans embryos. The antibodies have been used to
follow the granules, termed P granules, during the early embryonic
cleavage stages and throughout larval and adult development. P
granules become progressively localized to the germ-line precursor
cells during early embryogenesis. We are using conditionally lethal
maternal-effect mutations to study this localization process. In
addition to providing a rapid assay for P granules in wild-type,
mutant, and experimentally manipulated embryos, the antibodies also
promise to be useful in biochemically characterizing the granules and
in investigating their function.

   [*]From the Department of Molecular, Cellular, and Developmental
Biology; University of Colorado, Boulder, CO 80309.

INTRODUCTION

The question of how an egg establishes polarity and segregates different developmental potential to different embryonic blastomeres has intrigued developmental biologists for over a century (Wilson, 1925). An understanding of this process is fundamental to our understanding of development, and yet the process is as enigmatic today as it was to the embryologists of the 19th century. We still do not know the molecular basis of polarity, the nature of differentially partitioned determinants, and the mechanism of this asymmetric partitioning according to the polarity of the egg.

A well studied example of an asymmetrically segregating determinant is the special egg cytoplasm that is partitioned into the germ line. In many animals the germ line is set apart from the somatic lineages early in embryogenesis, and in most animals examined, there is a cytological manifestation of the establishment of the germ line (Eddy, 1975). The germ-line cytoplasm in embryos contains distinctive electron-dense structures that have been visualized by electron microscopy and are usually referred to as germ plasm, polar plasm, or polar granules. Morphologically similar structures, referred to as nuage, are seen attached to the nuclear envelope in gametes. Nuage and germ plasm are thought to represent a common component present in the germ-line cells throughout the life cycle (Mahowald, 1971a). Experimental studies of insects, amphibia, and nematodes have provided us with some understanding of the role of germ plasm in germ cell formation. Some of these findings will be discussed briefly here; several detailed reviews have been published within the last decade (Beams and Kessel, 1974; Eddy, 1975; Smith and Williams, 1975; Mahowald, 1977).

The results from experiments in which germ plasm has been destroyed, transplanted, or redistributed in early embryos suggest that germ plasm is required for germ-cell development. In both Drosophila (Geigy, 1931; Warn, 1972) and amphibia (Bounoure, 1939; Smith, 1966), ultra-violet irradiation of the egg cytoplasm containing the distinctive germ-line structures results in adults that lack germ cells. Transplantation of unirradiated germ plasm into UV-irradiated Drosophila eggs (Okada et al., 1974; Warn, 1975) or Rana pipiens eggs (Smith, 1966) restores the capacity of the eggs to form germ cells. Moreover, injection of extra Xenopus germ plasm into the vegetal cytoplasm of recipient Xenopus eggs causes the production of supernumerary primordial germ cells (Wakahara, 1978). The most compelling evidence for a determinative role of germ plasm comes from the transplantation experiments of Illmensee and Mahowald (1974, 1976) in which posterior cytoplasm that contained polar granules was transplanted from a wild-type Drosophila preblastoderm embryo to anterior or mid-ventral regions of genetically marked recipient embryos. Cells similar to pole cells formed at these ectopic sites and produced functional germ cells after transplantation to the posterior tip of a host embryo. Transplantation of oocyte cytoplasm containing polar granules also caused normally isomatic blastoderm nuclei to become incorporated into functional germ-line precursor cells (Illmensee, 1976). Thus, posterior polar plasm of oocytes and embryos is competent to induce germ-cell determination during embryogenesis.

In embryos of certain nematodes and insects the germ-line cells retain their full complement of chromosomes while somatic cells undergo chromosome diminution or elimination (see Beams and Kessel, 1974, for review), thus providing another distinction between the germ line and soma early in embryogenesis. Boveri (1910) studied early Ascaris eggs in which the germ plasm was distributed to more than one blastomere at the first cleavage, either as a result of dispermy or centrifugation of the eggs. The embryonic cells that acquired germ plasm did not undergo chromosome diminution, suggesting that this cytoplasm functions to prevent chromosome diminution.

The presence of distinctive structures in the germ cytoplasm and the effects of germ cytoplasm on nuclei in the germ cells and on germ-cell development provide a graphic demonstration of cytoplasmic localization of developmental information. However, the mechanism of localization of germ plasm and the nature and function of germ-line-specific structures remain unknown.

The segregation and nature of macromolecular developmental determinants may be studied using probes to tag, isolate, and characterize these molecules. Monoclonal antibody technology (Kohler and Milstein, 1975) provides a means of generating specific immunologic probes to developmental components that in the past have been difficult to study using conventional rabbit antisera. We describe here experiments with monoclonal antibody probes which recognize cytoplasmic granules that are unique to the germ line and that seem to correspond to the distinctive germ-line-specific structures seen by electron microscopy.

The organism we have used for these studies is the free-living soil nematode, Caenorhabditis elegans, a small, hermaphroditic, self-fertilizing worm with a simple anatomy (959 somatic cells in an adult hermaphrodite) and a short generation time (3 1/2 days at 25°C). Males, which arise spontaneously at low frequency by nondisjunction, can mate with hermaphrodites, resulting in production of outcross progeny in addition to self progeny. C. elegans is well suited to studying the segregation of developmental potential for the following reasons:

1. C. elegans embryos display visible asymmetry immediately after fertilization. Cleavage of the fertilized egg includes a series of four asymmetric stem cell divisions (Fig. 1). The daughter cells generated by these asymmetric

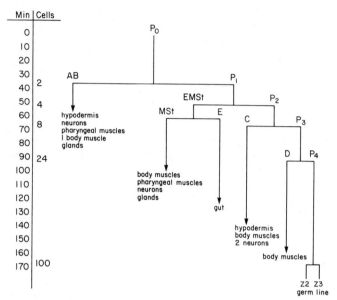

Figure 1: Lineage relationships, timing, and progeny cell fates in
the early cleavage divisions of C. elegans embryos at 25°C.
Progeny cell fates are based on data from Deppe et al. (1968),
Krieg et al., 1978), and Sulston et al. (1983).

cleavages differ in size, cell-cycle time, and developmental
potential; each cleavage gives rise to a larger somatic
blast cell and a smaller P cell. As a result, five somatic
lineages and the germ-cell (P) lineage are established by
the 16-cell stage of embryogenesis. The pattern of
asymmetric divisions and developmental fates is invariant
from embryo to embryo.

2.   The lineage of the entire organism, from fertilized egg to
     mature adult, has been determined by Sulston et al. (1983).
     Thus, for any cell in the organism, the ancestry and
     developmental fate(s) of its progeny are known.

3.   C. elegans is amenable to genetic analysis and manipulation
     (Brenner, 1974), and numerous mutants have been isolated and
     partially characterized. For the work described here, the
     existence of mutants with defects in gametogenesis and
     embryogenesis has proven extremely useful.

Immunologic Probes that Recognize Germ-Line Granules

Antibody binding to germ-line-specific granules was initially observed by immunofluorescence microscopy while screening hybridoma cell lines for production of monoclonal antibodies to C. elegans antigens. Fluorescein-conjugated rabbit anti-mouse IgG (F-RAM) is routinely used as a secondary antibody for detecting binding of mouse monoclonal antibodies to C. elegans tissues. An F-RAM preparation from Miles Laboratories was found to react with fixed C. elegans embryos directly, specifically staining cytoplasmic granules in the germ-line cells of embryos that had not been incubated with primary mouse antibody (Strome and Wood, 1982). Because these structures are unique to the germ-line or P lineage, we termed them P granules. Specific staining of P granules was observed using several antibody preparations from Miles, including both F-RAM and F-goat anti-rabbit (F-GAR), but not using F-RAM and F-GAR from other companies. One possible explanation for such an anti-P-granule antibody in the Miles preparations is that one or more of the immunized animals had a nematode infection that elicited production of antibodies cross-reactive to C. elegans antigens. Based on this possibility, serum samples from 26 rabbits from a local rabbit farm were screened for the presence of anti-P-granule antibodies by immunofluorescence microscopy; the sera from two rabbits contained such antibodies. Both the F-RAM and the two local rabbit sera specifically stain P granules in embryos. However, in later stage larvae and adults, these antibodies stain nongonadal as well as gonadal tissues, as if additional antigens recognized by the rabbit antibodies arise during larval development. This observation supported the hypothesis that these antibodies are present in the serum of unimmunized animals as a result of nematode infection and suggested that P granules might be highly antigenic. The presence of multiple antibody species in these sera also emphasized the need for more specific immunologic probes for P granules.

To obtain such probes, we used the rabbit anti-P-granule antibodies to optimize an immunofluorescence screen for monoclonal antibodies directed against P granules. Mice were immunized with a crude C. elegans early egg homogenate; following hyperimmunization, the mouse spleen cells were fused to mouse myeloma cells to establish hybridoma cell lines. Using the immunofluorescent screen, a hybridoma cell line was identified that secretes IgM directed against P granules. Therefore, P granules are evidently antigenic enough to elicit an immune response in the presence of many other nematode antigens, and the monoclonal technology has enabled us to prepare a specific probe to an antigen that probably represents only a minor component of the initial immunogen.

Thus we presently have two types of immunologic probes, rabbit sera and mouse monoclonal antibodies, with which to study P granules. All these anti-P-granule antibodies show the same patterns of

staining of fixed C. elegans embryos and dissected larval and adult
gonads. The monoclonal antibody is an excellent reagent for
immunofluorescent staining of P granules at all stages of C. elegans
development and does not result in extraneous staining of other
tissues.

The species specificity of the original F-RAM preparation was
tested using Drosophila melanogaster preblastoderm embryos, mouse
oocytes and cleavage stage embryos, Ascaris suum embryos, and embryos
from Panagrellus redivivus, a nematode closely related to C. elegans.
Under the conditions of fixation required for P-granule staining in
C. elegans embryos, Drosophila embryos showed no specific staining of
the posterior polar plasm or pole cells above a fairly high level of
background staining. Mouse oocytes and embryos and Ascaris embryos
also gave negative results. Only the Panagrellus embryos showed
specific staining of what are presumably the germ-line precursor
cells. The cross-reactivity of the monoclonal anti-P-granule
antibody has not yet been examined.

Behavior of P Granules During the Life Cycle

The segregation of P granules during embryogenesis in C. elegans
provides a graphic demonstration of the establishment of polarity and
cytoplasmic localization (Fig. 2). In the center of Fig. 2 is a
one-cell egg at metaphase of the first cleavage; the
anterior-posterior axis has been established, and P granules are
localized in the posterior cytoplasm which is destined for the P-cell
daughter. In the later stage embryos in Fig. 2, P granules are
observed in only the two germ-line precursor cells, as a result of
segregation at each of the early divisions.

Figure 3 shows the initial stages of P-granule coalescence and
segregation after fertilization. In the newly fertilized egg,
anti-P-granule antibodies stain numerous fine particles dispersed
throughout the cytoplasm (Fig. 3a). A dramatic change takes place
prior to the first cleavage. The particles begin to coalesce and
become localized in the posterior cytoplasm by the time of pronuclear
fusion (Fig. 3b). The granules are progressively segregated into the
smaller P cell at each of the early asymmetric divisions that
generate a somatic and a P-cell daughter (Fig. 3b), resulting in
granule localization to the P4 cell by the 16-cell stage of
embryogenesis.

We hypothesize that the presence of detectable P granules in
only the germ-line precursor cell of the 16-cell embryo may result
from two processes: movement of most of the granules into the
cytoplasm destined for the next P cell daughter at each of the early
asymmetric cleavages, and destruction of granules incorporated into a
somatic blastomere instead of a P cell. The movement of granules
into presumptive P-cell cytoplasm is suggested by the observation

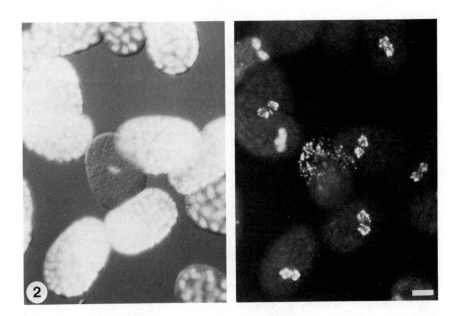

Figure 2: Immunofluorescence staining of P granules in early and
late embryos. Embryos in this and subsequent figures were fixed,
treated with a nonfluorescent primary antibody (in this case, mouse
monoclonal anti-P-granule antibody) followed by a
fluorescein-conjugated secondary antibody (fluorescein-conjugated
goat anti-mouse IgG), washed, stained with diamidinophenylindole
(DAPI) hydrochloride, rinsed, mounted for microscopy, and
photographed as described in Strome and Wood (1982). The left
panel shows the Nomarski differential interference contrast image
of the embryos with DAPI-stained chromosomes made visible by
fluorescence. The right panel shows the same embryos with
antibody-stained P granules made visible by fluorescence at a
different wavelength. The embryo in the center of the field is in
metaphase of the first cell cycle; it is oriented posterior-up,
anterior-down. Note the localization of P granules in the
posterior cytoplasm. Most of the later stage embryos in the field
show P-granule staining in only the two germ-line precursor cells.
Bar = 10 um.

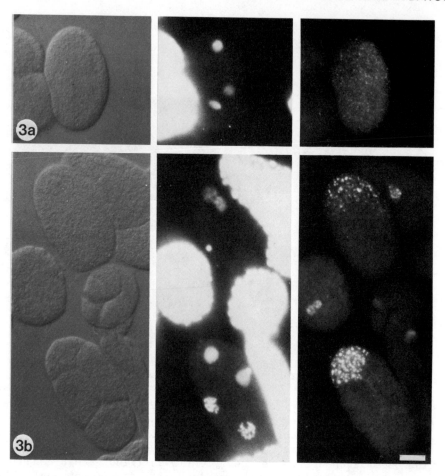

<u>Figure 3</u>: Localization of P granules after fertilization. Embryos
were prepared and photographed as in Fig. 2. The panels from left
to right show the Nomarski, DAPI, and immunofluorescence images,
respectively. The orientation is posterior-up, anterior-down. A)
One-cell embryo shortly after fertilization. The male and female
pronuclei are still at opposite ends of the embryo; P granules are
dispersed throughout the cytoplasm. B) One-cell embryo at the time
of pronuclear fusion (uppermost embryo) showing prelocalization of
P granules; four-cell embryo (lower embryo) showing the presence of
P granules in only the P2 cell. Bar = 10 um.

that most or all of the detectable P granules are asymmetrically
distributed as early as prophase of each cell cycle (Fig. 4).
However, such an asymmetric distribution could also be due to
preferential degradation during each cell cycle of P granules in the
region of the cytoplasm destined for the non-P somatic cell. Such a

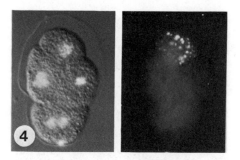

Figure 4:   A four-cell embryo showing prelocalization of the P
  granules prior to cleavage. The left panel shows the Nomarski/DAPI
  image with posterior-up, anterior-down; the P2 nucleus in the upper
  cell is in prophase. The right panel shows the immunofluorescence
  staining of P granules which are prelocalized in the ventral region
  of the P2 cells. Bar = 10 um.

mechanism for destruction of P granules that are distributed to
somatic blastomere cytoplasm seems to exist. The partitioning of P
granules to the P-cell lineage is not absolute, and small granules
are sometimes observed in the somatic sister of a P cell (Fig. 5).
However, these small granules apparently usually do not persist for
longer than one additional cell cycle because they are very rarely
detected in the progeny of the somatic cell, as if P granules are
unstable in the somatic cytoplasm.

    The size, number, and distribution of P granules change markedly
during the early embryonic cleavage period. In one-cell to four-cell
embryos, they are numerous (3-4 dozen) and small, located apparently
randomly in the cytoplasm during interphase and near the cortex
during cell division (Figs. 2, 3). However, in older embryos the
small granules have coalesced into 3-5 large structures that are
located around the nucleus of the P cell (Fig. 6).

    Cytoplasmic inclusions characteristic of germ plasm have also
been observed in cells of the P lineage by transmission electron
microscopy of sectioned C. elegans embryos (Wolf et al., 1983).
These inclusions, like P granules, are restricted to the P cells
during the early cleavage stages. Moreover, their number, size, and
distribution change in the same manner and at the same times as
observed for P granules. These correlations suggest that the
anti-P-granule antibodies react with a component(s) of the germ
plasm.

    Antibody staining of embryos, larvae, and adults indicates that
P-granule antigens are present throughout the life cycle of the worm
(Strome and Wood, 1982). In a newly hatched larva the gonad

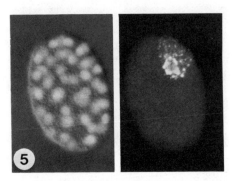

Figure 5: P granules in the somatic sister as well as the P cell.
The left panel shows the Nomarski/DAPI image and the right panel
the immunofluorescence image. Posterior is up; anterior is down.
In addition to P granules in the P4 cell, small granules are
detectable in the two D cells. Bar = 10 um.

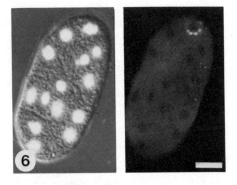

Figure 6: Perinuclear P granules in the P3 cell of a 15-cell embryo.
The left panel shows the Nomarski/DAPI image and the right panel
the immunofluorescence image. Posterior is up; anterior is down.
The numerous small granules present in early embryos have coalesced
into several large perinuclear structures. This embryo was stained
with F-RAM from Miles (Strome and Wood, 1982). Bar = 10 um.

primordium consists of four cells: Z1, Z2, Z3, and Z4. Z2 and Z3
are derived from the embryonic P4 cell and give rise to all of the
germ cells in the adult worm; Z1 and Z3 are derived from the MSt
lineage and give rise to the somatic gonad. In the gonad primordium
of a young larva, anti-P-granule antibodies stain perinuclear
granules in Z2 and Z3 (Fig. 7); no staining is detectable in Z1 and
Z4. During larval development, staining is confined to the
descendants of Z2 and Z3. The staining becomes fainter as the germ
cells proliferate, suggesting that the granules originally present in

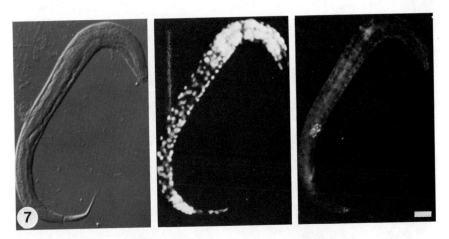

Figure 7: Antibody staining of P granules in a newly hatched larva.
Nomarski, DAPI, and immunofluorescence images (left, middle, and
right, respectively) of a first-stage larva, showing staining of
only the Z2 and Z3 cells in a four-cell gonad primordium (evident
in the Nomarski image). The two somatic gonad precursor cells, Z1
and Z4, do not stain for P granules. This larva was stained with
F-RAM from Miles (Strome and Wood, 1982). Bar = 10 um.

Z2 and Z3 are distributed among their progeny cells. In late larval
and adult gonads, staining of perinuclear granules is again intense
(Fig. 8), suggesting that new P granules are synthesized during
gametogenesis in older larvae and adults.

The behavior of P granules differs during oogenesis and
spermatogenesis. In adult hermaphrodites, each half of the
symmetrical gonad is a U-shaped structure consisting of an ovary in
the distal arm and an oviduct in the proximal arm, connected through
a spermatheca to the common uterus. The progression of germ-cell
differentiation shows a characteristic spatial organization (Hirsh et
al., 1976; see Fig. 8). Germ cells divide mitotically in the distal
tip of the overy and enter meiosis in the more proximal region of the
distal arm. Late in gonadogenesis, these cells give rise to about
150 sperm in each arm of the gonad. Sperm production then ceases and
oogenesis commences, continuing throughout the reproductive life of
the hermaphrodite. Oocytes form at the bend of the gonad, mature in
the proximal arm, and subsequently are fertilized in the spermatheca.
In the distal arm of the adult hermaphrodite gonad, P granules are
perinuclear and stain intensely (Fig. 8). During oogenesis the
granules remain associated with nuclei as the cells enter meiosis.
As oocytes mature, the granules disaggregate and disperse from their
perinuclear location (Fig. 8). Mature oocytes show staining of small
particles randomly distributed in the cytoplasm; the granules remain

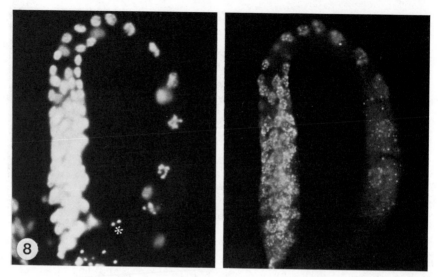

Figure 8: Staining of P granules in the distal and proximal arms of
a dissected adult hermaphrodite gonad. The DAPI image on the left
shows the progression of germ cell differentiation. The
mitotically dividing germ nuclei of the distal arm are on the left
side of the picture. The germ cells enter meiosis just prior to
the bend of the gonad. Oocytes form at the bend and mature in the
proximal arm (right side of the picture). The immunofluorescence
image on the right panel shows that the perinuclear granules
disperse during oocyte maturation; they appear randomly distributed
in the cytoplasm of mature oocytes. The asterisk marks the
location of two sperm, characterized by their highly condensed
nuclei. They are not stained by the monoclonal anti-P-granule
antibody.

in this dispersed state until after fertilization (Fig. 3) when the
cycle of P-granule coalescence and segregation begins again.

     C. elegans sperm differ from the flagellated sperm of many
invertebrates and vertebrates; they are amoeboid, and because they
contain a significant volume of cytoplasm, they can contribute
cytoplasmic components at fertilization. However, P granules in the
embryo seem to be contributed entirely by the oocyte because mature
sperm obtained from either males or hermaphrodites lack P granules as
detected by immunofluorescence using the monoclonal antibody (Fig.
8).

     The gonads of adult males are similar in spatial organization to
hermaphrodite gonads (Wolf et al., 1978). Male gonads are
single-armed asymmetric organs in which germ-cell nuclei divide
mitotically in the distal arm and enter and complete meiosis in the

proximal arm. When treated with the monoclonal antibody, the distal
region of male gonads displays the same intense perinuclear staining
as observed in hermaphrodite gonads, but staining disappears from
late pachytene germ cells (Fig. 9), and spermatocytes and mature
sperm are not stained above background levels. The original F-RAM
preparation from Miles and the two rabbit sera that react with P
granules do show cytoplasmic staining of sperm. However, the
staining patterns of the three rabbit antibodies differ and probably
involve reactions with antigens other than P granules. As mentioned
previously, the three rabbit antibody preparations are specific for P
granules in embryos but stain other tissues as well in later-stage
larvae and adults.

So far we can only speculate about the relation of P granule
behavior to possible functions. During oogenesis they may disperse
from the nuclear membrane so that a cytoplasmic segregation mechanism
can partition them to the germ-cell lineage. Alternatively, this
dispersion, occurring as both oocytes and sperm proceed through
meiosis, may indicate that P granules function during germ-cell
differentiation. It is not known whether P granules are required for
either determination or differentiation of the germ line in C.
elegans. Only in Drosophila is there unequivocal evidence that
cytoplasm containing polar granules is determinative for the germ
line (Illmensee and Mahowald, 1974, 1976; Illmensee, 1976), and even
in this case the role of the polar granules themselves is not known.

The striking similarity between C. elegans germ granules and
Drosophila polar granules is worth noting. By electron microscopy
the granules in both organisms appear as amorphous masses of
fibrillar material, lacking surrounding membranes, and measuring
approximately 0.5-1 micron in diameter in early embryos (Mahowald,
1968, 1977; Wolf et al., 1983). In both organisms the granules are
associated with mitochondria early in embryogenesis. In Drosophila
the polar granules lose this association during the early embryonic
cleavage stages and appear surrounded by ribosomes. Association of
the germ-cell granules with free ribosomes has not been detected in
C. elegans embryos, although the granules are often found close to
endoplasmic reticulum (Wolf et al., 1983). In both organisms the
germ granules change in the degree of aggregation with each other
during the early embryonic period and eventually assume perinuclear
locations in the germ-line precursor cells. The continuity of polar
granules throughout the life cycle of Drosophila has been inferred
from observation of similar electron-dense material in the germ cells
at almost all stages of maturation (Mahowald, 1971a). By
immunofluorescence, P granules appear to be present throughout the
life cycle of C. elegans. In Drosophila the possibility that polar
granules are self-replicating cytoplasmically inherited components
was tested by making interspecific nucleo-cytoplasmic hybrid pole
cells. In this way, Mahowald et al. (1976) showed that the polar
granules are not cytoplasmically inherited but depend for their

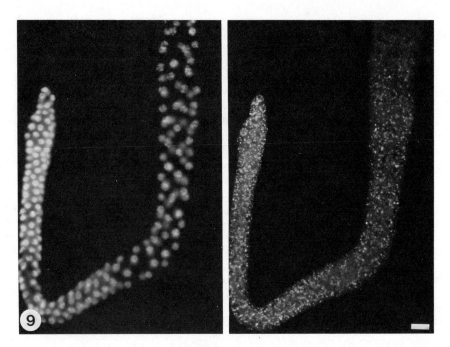

Figure 9: Staining of P granules in a dissected adult male gonad. The DAPI image on the left shows the progression of germ cell differentiation. The immunofluorescence image in the right panel shows the P granules surrounding the mitotically dividing germ nuclei in the distal arm (left side of each picture) and the pachytene nuclei in the bend of the gonad. The monoclonal anti-P-granule antibody does not stain spermatocytes (upper right side of each picture) or mature sperm (not shown in this picture). Bar = 10 um.

formation on expression of the nuclear genome during oogenesis. The requirement for nuclear genome expression for P-granule regeneration in C. elegans has not yet been examined. Perhaps the most striking contrast between Drosophila and C. elegans is that the localization of Drosophila polar granules in the posterior cytoplasm of oocytes occurs prior to fertilization (stage 9 of oogenesis), while in C. elegans, P granules do not become localized until after fertilization, and segregation occurs at each of the early embryonic cell divisions.

How Are P Granules Segregated During Early Embryonic Cleavage?

The localization of P granules in early C. elegans embryos provides documentation of asymmetry in the cytoplasm of early blastomeres. We are using mutants in gametogenesis and conditionally lethal maternal-effect mutants in early embryogenesis, some of which appear to be defective in the establishment of embryonic polarity, to try to understand the molecular basis for polarization and the mechanism by which P granules are segregated in response to the polarity of the egg. Although the results of such analyses must be interpreted with caution because the primary defects caused by the mutations are not well understood, the mutants are providing useful information and testable hypotheses.

Mutants defective in sperm function were used to ask the question of whether P-granule segregation is dependent upon fertilization or, alternatively, whether the oocyte can direct coalescence and segregation in the absence of fertilization. The temperature-sensitive allele b245 at the ixs-2 locus prevents sperm production and hence converts hermaphrodite worms into "true" females at the restrictive temperature (Edgar, 1982). Two other temperature-sensitive alleles, hc1 and hc24 at the fer-1 locus, result in the production of defective sperm at the restrictive temperature (Ward and Carrel, 1979). In these mutant animals the oocytes that pass through spermathecae containing no sperm or defective sperm are not fertilized; they undergo DNA endoreplication, but no mitosis or cleavage. Oocytes that come into contact with defective fer-1(hc1) sperm also show the increased Brownian and saltatory motion of cytoplasmic granules that is normally seen after fertilization and has been termed "cytoplasmic activation." In both isx-1(b245) and fer-1(hc1 or hc24) oocytes, coalescence of P granules does occur after passage through the spermatheca, but segregation of the granules to one end of the egg does not (Fig. 10; Table 1). Thus the oocyte apparently can initiate P-granule coalescence, perhaps triggered by germinal vesicle breakdown. However, the oocyte does not seem to contain or is not able to utilize the positional information necessary for granule segregation to one pole, even when the "cytoplasmic activation" characteristic of fertilized eggs occurs.

One possible mechanism for P-granule segregation would be via attachment of the granules to the aster destined for the P cell, although the prelocalization of P granules in the presumptive P-cell cytoplasm prior to spindle formation would argue against such a mechanism. Mutants with aberrant early cleavage patterns were used to ask whether the position and orientation of the mitotic spindle and cleavage furrow are important for granule segregation. A wild-type zygote divides transversely and asymmetrically, distributing P granules to the smaller posterior P cell. Mutations at the zyg-9 locus (for example, the temperature-sensitive allele b244), cause zygotes from mutant hermaphrodites grown at the restrictive temperature to undergo a longitudinal first cleavage in which the spindle is rotated 90° from its normal position (Wood et al., 1980). Despite the altered orientation of the spindle and cleavage furrow, P granules become localized at the posterior pole of the embryo and thus are distributed to both of the resulting

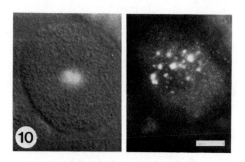

Figure 10:   Immunofluorescence staining of P granules in an
  isx-1(b245) oocyte. The left panel is the Nomarski/DAPI image and
  the right panel the immunofluorescence image. The granules have
  coalesced in the central region of the oocyte. Bar = 10 um.

blastomeres (Fig. 11a; Table 1). This result suggests that P
granules do not segregate either with the microtubules of one
half-spindle or with one nucleus and that the positional information
responsible for P-granule segregation is independent of spindle
orientation. At the second (transverse) cleavage of zyg-9(b244)
embryos, P granules are asymmetrically segregated, resulting in
granules in two of the four to five blastomeres present at this stage
(Fig. 11b). This altered distribution of P granules into more than
one early blastomere is reminiscent of the experiments of Boveri
(1910) in which germ plasm was distributed to multiple cells in
Ascaris embryos after tetrapolar cleavage of a dispermic zygote.
Germ plasm became localized in the vegetal cytoplasm despite abnormal
spindle orientation. In embryos containing germ plasm in multiple
cells due to either fertilization by two sperm or centrifugation of
normal zygotes, the presence of germ plasm was correlated with
prevention of chromosome diminution. Since C. elegans embryos do not
undergo chromosome diminution, and since zyg-9(b244) embryos arrest
later in embryogenesis, we cannot determine whether redistribution of
P granules to more than one early blastomere results in the
production of supernumerary germ cells.

     Several mutant alleles at the zyg-11 locus cause zygotes from
mutant hermaphrodites to exhibit variable positioning of the first
cleavage plane, as shown in Table 1 (Wood et al., 1980). The
localization of P granules in zyg-11 mutant embryos is also variable
and difficult to interpret. In a strain carrying zyg-11(mn40), a
probable null allele (unpublished experiments), a large percentage of
embryos displaying each of the three different cleavage patterns have
P granules localized in the end of the embryo next to the polar body
which, in wild-type embryos, is almost always anterior (Fig. 12;
Table 1). It is not known whether in these mutant embryos the
polarity of the embryo is reversed relative to the polar body or
whether P granules are segregated to the wrong end of the embryo. In

Table 1. Summary of Phenotypes and Distributions of P Granules in
Mutant C. elegans Embryos

Locus (Allele)	Phenotype	Distribution of P granules
wild-type		a
isx-2(b245)	no sperm	
fer-1(hcl, hc24)	defective sperm	
zyg-9(b244)	first cleavage plane perpendicular to wild-type orientation	
zyg-11(b2, mn40)	variable position of first cleavage plane	b
emb-27(g48)	no cleavage	

a Reference wild-type embryo, oriented anterior-left, posterior-right. The
  polar body is adjacent to the anterior AB cell. Nuclei are represented as
  open circles in each blastomere; P granules are represented as small dots,
  in this case in the posterior P1 cell.

b In one experiment in which 76 1-cell to 4-cell mn40 mutant embryos were
  examined, 44 (58%) showed P granules at the end of the embryo next to the
  polar body. In the remaining embryos, granules were located in the center
  of the embryo, at the end of the embryo opposite the polar body, or throughout the embryo.

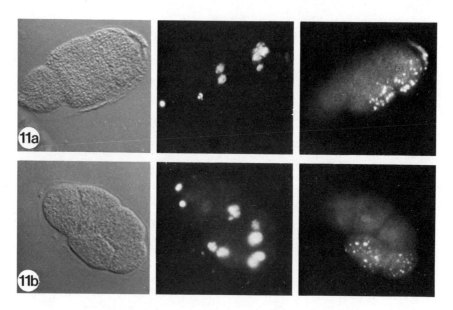

Figure 11:   Immunofluorescence staining of P granules in zyg-9(b244)
embryos.   Nomarski (left panel) DAPI (center panel), and
immunofluorescence (right panel) images are shown for a two-cell
(a) and a four-cell (b) embryo.   Anterior is left; posterior is
right.   The embryo in 'a' underwent an abnormal longitudinal first
cleavage to yield an anterior cytoplast and two elongated cells of
about equal size, which rotated before fixation so that one cell
lies posterior to the other.   Both cells contain P granules which
became localized normally at the posterior pole of the zygote
despite the abnormal transverse axis of the first-cleavage spindle.
The embryo in 'b' underwent a similar first cleavage followed by a
second asymmetric cleavage in which the P granules were segregated
to the two posterior daughter cells.   Bar = 10 um.

many of the symmetrical two-cell and four-cell zyg-11 mutant embryos,
P granules are found in all of the cells.   These results support the
suggestion that this mutant is defective in the establishment of
polarity (N. Wolf and K. Kemphues, personal communication).

Mutants in which early cleavages are blocked were used to ask
about the role cleavage plays in partitioning P granules.   A
temperature-sensitive allele at the emb-27 locus, g48, prevents
cleavage of zygotes from mutant hermaphrodites reared at the
restrictive temperature (Denich et al., 1983).   Mutant embryos
undergo several rounds of mitosis but no cytokinesis.   P granules
coalesce in these mutant embryos but do not become localized at one
pole; instead they aggregate in the center of the embryo (Fig. 13;
Table 1).   One possible explanation for these results is that the

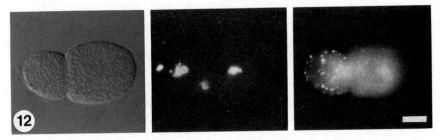

Figure 12: Antibody staining of P granules in a zyg-11(mn40) embryo.
The panels from left to right show Nomarski, DAPI, and
immunofluorescence images, respectively, of a two-cell embryo that
divided transversely and asymmetrically, giving rise to a smaller
cell adjacent to the polar body (left of this picture). P granules
are localized in the cell next to the polar body which, in
wild-type embryos, is almost always anterior. Bar = 10 um.

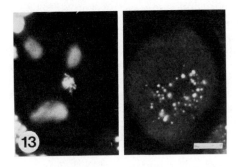

Figure 13: Immunofluorescence staining of P granules in an
emb-27(g48) embryo. The left panel is the DAPI image and the right
panel the immunofluorescence image. The granules have coalesced in
the central region of the embryo. Bar = 10 um.

cleavage defects in emb-27(g48) embryos reflect a defect in
microfilament-mediated processes or processes that involve the
cortex, and that microfilaments or the cortex are involved in
P-granule partitioning. In wild-type embryos, P granules become
localized at the cortex in preparation for partitioning into the next
P cell; the granules lose this cortical association and are not
asymmetrically localized in emb-27(g48) embryos. Consistent with a
role of microfilaments or cortical actin in granule movement is the
effect on C. elegans embryos of cytochalasin B, an inhibitor of
microfilament-mediated processes. C. elegans embryos that are
rendered permeable to cytochalasin B display the same phenotype as
emb-27(g48) embryos. They undergo rounds of mitosis but not
cytokinesis; P granules collapse into the center of the

cleavage-blocked P blastomeres and are no longer segregated during
subsequent rounds of mitosis. Thus, both mutational and drug-induced
manipulations of cleavage and related processes in C. elegans embryos
suggest an involvement of microfilaments or the cortex or both in
P-granule movement.

FUTURE PROSPECTS

The anti-P-granule antibodies described here provide a promising
approach to studying germ-cell-specific elements. All previous
studies on nuage and germ plasm have relied on electron microscopy as
an assay. These studies, while informative, are laborious, and
inferred relationships between different forms of germ plasm at
different stages of the life cycle have been based solely on
morphological similarities. The availability of immunologic probes
may allow us to bridge the gap between the morphology and the
biochemistry and function of germ-cell structures. Some of the
questions that may be addressed using the antibodies are the
following:

1. Do P granules correspond to nuage or germ plasm seen in C.
    elegans by electron microscopy? The strict correlation at
    every stage examined between the number, size, and location
    of the structures seen by electron microscopy and P granules
    visualized by immunofluorescence suggests that they
    correspond. Because fixation methods for electron
    microscopy and immuno-staining with anti-P-granule
    antibodies are incompatible, direct tests of this suggestion
    have not yet been carried out. However, microinjection of
    anti-P-granule antibody into C. elegans blastomeres,
    followed by fixation of the embryos for electron microscopy
    and binding of secondary antibody conjugated to an electron
    dense marker, should allow simultaneous visualization of P
    granules and nuage.

2. What are the biochemical constituents of P granules and do
    they include RNA? The protein components of P granules may
    be identified by gel blotting techniques or
    immunoprecipitation or both. The latter approach, which
    involves precipitation of antibody bound to radiolabeled
    antigen followed by polyacrylamide gel analysis of the
    labeled polypeptide(s), should provide a powerful means of
    identifying the antigen(s) recognized by each antibody
    preparation and other macromolecular constituents associated
    with the antigen(s) in intact P granules.

3. Do P granules play a role in determination, differentiation,
    or function of germ cells? It is not known whether P

granules are required for development of germ cells in C. elegans, but several approaches may be used to test such a requirement. Microinjection of anti-P-granule antibodies into adult gonads or embryonic P cells may provide a means of functionally inactivating P granules, either by antibody binding to critical sites or by formation of microimmunoprecipitates. Alternatively, it may be possible to functionally inactivate P granules by UV irradiation of the cytoplasm containing them in C. elegans embryos, as germ plasm has been inactivated by UV irradiation in Drosophila and Xenopus embryos (Geigy, 1981; Warn, 1972; Bounoure, 1939; Smith, 1966). Of particular promise is the possibility of correlating the state of the granules, assayed using the antibodies, with an effect of irradiation such as induced sterility of the worms obtained from irradiated embryos.

Laufer and von Ehrenstein (1981) devised a method of extruding cytoplasm from early C. elegans embryos through laser-produced holes in the eggshell. Many of the embryos lacking a significant quantity of posterior cytoplasm continued to develop and gave rise to healthy, fertile adults. These experiments suggested that development of this organism may not depend on prelocalization of determinants in specific regions of the egg cytoplasm. However, an alternative possibility is that developmental determinants, if they exist, are anchored to the cytoskeleton or cortex and are not free to be extruded. Staining of both the extruded fragments and the treated embryos and larvae with anti-P-granule antibody should clarify whether P granules, which represent a clear example of prelocalized cytoplasmic components, are extruded. If they can be extruded without deleterious effect on future germ cells, then they may be neither determinative nor required for germ-line function.

Lastly, a genetic approach to examination of the role of P granules may be fruitful. It may be possible to isolate mutants defective in P granule synthesis, formation, or segregation. The immunofluorescence studies of P granules indicate that they are contributed to the embryo through the oocyte. If P granules from the oocyte function to determine germ cells in the next generation, then such a homozygous mutant hermaphrodite might show the grandchildless phenotype; that is, give rise to sterile offspring. This phenotype has not been observed in previous mutagenesis experiments with C. elegans, but more extensive screening for grandchildless mutants would be worthwhile. As described in the chapter by Konrad and Mahowald, several alleles at a locus called tudor now exist in Drosophila melanogaster. These mutants show a maternal effect and result in a grandchildless phenotype as well as some other defects. The exciting finding that polar granules are defective in the embryos that give rise to sterile adults supports a role for the polar granules themselves in germ-cell determination.

CONCLUDING REMARKS

Germ cells in a wide variety of invertebrate and vertebrate species contain unique organelles that are present in the cytoplasm of the oocyte and are passed on to the next generation of germ cells via a process of cytoplasmic localization. The ubiquity of such structures has led to speculation that they play an important role in determination or differentiation of the germ line. Such a role of the cytoplasm containing these structures in germ-cell determination has been borne out in the case of Drosophila. However, the biochemical nature of the organelles and their function in germ-cell development remain mysterious, as does the mechanism by which the embryo insures its transmission from one generation of germ cells to the next. We have developed immunofluorescence techniques for visualizing germ-line-specific granules in C. elegans that are probably the same as those seen by electron microscopy. By the criterion of reaction with a monoclonal antibody, at least one component of these granules is continuous throughout the life cycle despite changes in the morphology and distribution of granules within cells. Preliminary results suggest that the asymmetric segregation of these granules during the early embryonic cleavages is effected not by association with mitotic spindles, but rather by a microfilament-based mechanism. The availability of immunologic probes directed against these granules will facilitate further study of their identity, mechanism of segregation, biochemical composition, and perhaps function.

ACKNOWLEDGEMENTS

We would like to thank Dorothy Cohen for technical assistance with tissue culture. This research was supported by postdoctoral fellowships to S.S. from the Anna Fuller Fund and National Institutes of Health, and by grants to W.B.W. and S.S. from the National Institutes of Health (HD14958) and the American Cancer Society (CD-96).

REFERENCES

Beams, W.H. and Kessel, R.G. (1974). The problem of germ cell determinants. Inter. Rev. Cytol. 39:413-479.

Bounoure, L. (1939). L'Origine des Cellules Reproductrices et la Probleme de la Ligne Germinale. Gauthier-Villars, Paris.

Boveri, T. (1910). Uber die Tielung centrifugierter Eier von Ascaris megalocephala. Arch. Entwicklungsmechanik 30:101-125.

Brenner, S. (1974). The genetics of Caenorhabditis elegans. Genetics 77:71-94.

Davidson, E.H. (1976). Gene Activity in Early Development, Chapter 7. Academic Press, New York.

Denich, K., Isnenghi, E., Cassada, R., and Schierenberg, E. (1983). Cell lineages and developmental defects of temperature-sensitive arrest mutants of the nematode Caenorhabditis elegans. Dev. Biol. (in press).

Deppe, U., Schierenberg, E., Cole, T., Krieg, C., Schmitt, D., Yoder, B., and von Ehrenstein, G. (1978). Cell lineages of the embryo of the nematode Caenorhabditis elegans. Proc. Natl. Acad. Sci. USA 75:376-380.

Eddy, E.M. (1975). Germ plasm and the differentiation of the germ cell line, G.H. Bourne and J.F., Danielli (eds.). Int. Rev. Cyt. 43:2290-280.

Edgar, L. (1982). Control of spermatogenesis in the nematode Caenorhabditis elegans. Ph.D. thesis, University of Colorado.

Geigy, R. (1931). Action de l'ultra-violet sur le pole germinal dans l'oeuf de Drosophila melanogaster. Rev. Suisse Zool. 38:187-288.

Hirsh, D., Oppenheim, D., and Klass, M. (1976). Development of the reproductive system of Caenorhabditis elegans. Dev. Biol. 49:200-219.

Illmensee, K. (1976). Nuclear and cytoplasmic transplantation in Drosophila. Symposium on Insect Development, P. Lawrence (ed.). London.

Illmensee, K. and Mahowald, A.P. (1974). Transplantation of posterior polar plasm in Drosophila. Induction of germ cells at the anterior pole of the egg. Proc. Natl. Acad. Sci. USA 71:1016-1020.

Illmensee, K. and Mahowald, A.P. (1976). The autonomous function of germ plasm in a somatic region of the Drosophila egg. Exp. Cell Res. 97:127-140.

Kohler, G. and Milstein, C. (1975). Continuous cultures of fused cells secreting antibody of predefined specificity. Nature 256:495-497.

Krieg, C., Cole, T., Deppe, U., Schierenberg, E., Schmitt, D., Yoder, B., and von Ehrenstein, G. (1978). The cellular anatomy of embryos of the nematode Caenorhabditis elegans; analysis and reconstruction of serial section electron micrographs. Dev. Biol. 65:193-215.

Laufer, J.S. and von Ehrenstein, G. (1981). Nematode development after removal of egg cytoplasm: absence of localized unbound determinants. Science 211:402-405.

Mahowald, A.P. (1968). Polar granules of Drosophila. Ultrastructural changes during early embryogenesis. J. Exp. Zool. 167:237-262.

Mahowald, A.P. (1971). Polar granules of Drosophila. The continuity of polar granules during the life cycle of Drosophila. J. Exp. Zool. 176:329-344.

Mahowald, A.P. (1977). The germ plasm of Drosophila: an experimental system for the analysis of determination. Amer. Zool. 17:551-563.

Mahowald, A.P., Illmensee, K., and Turner, F.R. (1976). Interspecific transplantation of polar plasm between Drosophila embryos. J. Cell Biol. 70:358-373.

Okada, M., Kleinman, I.A., and Schneiderman, H.A. (1974). Restoration of fertility in sterilized Drosophila eggs by transplantation of polar cytoplasm. Dev. Biol. 37:43-54.

Smith, L.D. (1966). The role of a "germinal plasm" in the formation of primordial germ cells in Rana pipiens. Dev. Biol. 14:330-347.

Smith, L.D. and Williams, M.A. (1975). Germinal plasm and determination of primordial germ cells. The Developmental Biology of Reproduction, C.L. Market and J. Papaconstantinou (eds.), pp. 3-24. Academic Press, New York.

Strome, S. and Wood, W.B. (1982). Immunofluorescence visualization of germ-line-specific cytoplasmic granules in embryos, larvae, and adults of Caenorhabditis elegans. Proc. Natl. Acad. Sci. USA 79:1558-1562.

Sulston, J., Schierenberg, E., White, J., Thomson, N., and von Ehrenstein, G. (1983). The embryonic cell lineage of the nematode Caenorhabditis elegans. Dev. Biol. (submitted for publication).

Wakahara, M. (1978). Induction of supernumerary primoridal germ cells by injecting vegetal pole cytoplasm into Xenopus eggs. J. Exp. Zool. 203:159-164.

Ward, S. and Carrel, J.S. (1979). Fertilization and sperm competition in the nematode Caenorhabditis elegans. Dev. Biol. 73:304-321.

Warn, R. (1972). Manipulation of the pole plasm of Drosophila melanogaster. Acta Embryol. Exper. Suppl. 415-427.

Warn, R. (1975). Restoration of the capacity to form pole cells in UV-irradiated Drosophila embryos. J. Embryol. Exp. Morph. 33:1003-1011.

Wilson, E.B. (1925). The Cell in Development and Heredity. McMillan, New York.

Wolf, N., Hirsh, D., and McIntosh, J.R. (1978). Spermatogenesis in males of the free-living nematode Caenorhabditis elegans. J. Ultrastruc. Res. 63:155-169.

Wolf, N., Priess, J., and Hirsh, D. (1983). Segregation of germ-line granules in early embryos of Caenorhabditis elegans: an electron microscopic analysis. J. Embryol. Exp. Morph. (in press).

Wood, W.B., Heckt, R., Carr, S., Vanderslice, R., Wolf, N., and Hirsh, D. (1980). Parental effects and phenotypic characterization of mutations that affect early development in Caenorhabditis elegans. Dev. Biol. 74:446-469.

GENETIC AND DEVELOPMENTAL APPROACHES TO UNDERSTANDING

DETERMINATION IN EARLY DEVELOPMENT

Kenneth D. Konrad and Anthony P. Mahowald[*]

ABSTRACT

   In this paper we review experimental evidence relating to the
stage of determination of blastoderm cells of the Drosophila
melanogaster embryo. Cellular ablation studies and clonal analysis
reveal that the ectodermal precursors of both the larva and adult are
determined for segmental identity at blastoderm. The adult
ectodermal precursors, however, are not determined for specific
structures within segments. This situation is not clear with respect
to larval epidermal precursors. In contrast, cellular
transplantation studies indicate that mesodermal and endodermal
precursors of both the larva and adult are undetermined with respect
to segmental identity or tissue specificity at blastoderm. We also
review the maternal influence on cellular determination as revealed
by maternal effect mutations. Many of these mutations exert an
effect on the developmental fate of most or all of the blastoderm
cells. These global effects may be produced through a disruption of
gradients of maternal determinative gene products. Observations from
a developmental analysis of a maternal effect mutation disrupting
dorsal-ventral polarity reveals that maternal information may also be
involved in the cell interactions required for gastrulation. Other
maternal effect mutations appear to affect the developmental fate of
a more limited portion of the blastoderm surface. Maternal effect
mutations of either class can exhibit a strict maternal effect or a
varying degree of rescuability by the paternal genome.

   [*]From the Developmental Biology Center, Department of
Developmental Genetics and Anatomy, Case Western Reserve University,
Cleveland, OH 44106.

167

INTRODUCTION

One of the central unanswered questions of contemporary developmental biology is the molecular mechanism for establishing the determined state. Embryologists at the turn of the century demonstrated that the developmental potential of many cells of the early embryo is restricted to producing only portions of the future organism. The most convincing demonstrations of determination have depended upon the ability to manipulate the embryo in such a way that a specific cell or group of cells is challenged to differentiate either alone or in a foreign situation. In these instances, if the cell or cells continue to develop according to the fate they would have had in the embryo, they are considered determined for this fate. In other words, the cells have become restricted in their developmental potential such that only a portion of the cell fates of the organism are open to them.

The contemporary paradigm for understanding this phenomenon has evolved around the concept of cytoplasmic localization. As the nuclei in the embryo are exposed to different cytoplasmic environments they are thought to become committed to various programs of differential gene expression characteristic of differentiated cell types (reviewed by Davidson, 1976). This view assigns the contents and organization of the egg cytoplasm a position of primary importance in the determination of early embryonic cells. Virtually nothing, however, is known about the molecular nature of the signals to which embryonic nuclei respond or the mechanisms by which these signals become distributed during oogenesis and early development.

One of the most promising approaches to this problem has been the genetic analysis of early development. Maternal effect mutations disrupting early development have been identified in several organisms. The sophisticated genetics developed for Drosophila melanogaster has allowed a large number of such mutations to be produced in this organism. It is anticipated that current analyses of these mutations will soon be yielding insights into the molecular nature of determinative information in the Drosophila embryo. In addition, micromanipulation techniques have recently allowed the application of cellular transplantation, a direct test of cell determination, to the Drosophila embryo.

Although the cells of the Drosophila embryo are generally considered to be mosaic (Anderson, 1966), they actually appear to utilize a variety of mechanisms to achieve the determined state. One of the most convincing examples of specific determinants being localized during oogenesis, namely the germ plasm, has been analyzed extensively in Drosophila (Mahowald, 1977). Progress concerning this aspect of determination has been reviewed elsewhere (Mahowald and Boswell, 1983), and will not be discussed here. In this paper we will review the evidence in Drosophila for determination at the

blastoderm stage of somatic cell types. We will argue that
information conferring segmental identity on ectodermal derivatives
is progressively localized after fertilization but that determination
of all of the cell types within each segmental anlagen is incomplete
at cellularization. In particular, the mesodermal and endodermal
lineages do not appear to be determined at cellularization and may
become organized through inductive interactions with the ectoderm.
We will also review information from a number of maternal effect
mutations which indicate that the basis for the proper establishment
of the embryo pattern is dependent upon ooplasmic information which
interacts with the zygotic genome.

Experimental Approaches to Determination in Drosophila

Because of the syncytial nature of early Drosophila development
it is difficult to achieve an experimental demonstration of mosaicism
prior to cellularization. Elegant ligation experiments, however,
have provided us with some indication of the degree of mosaicism
built into the egg structure itself. The most extensive results on
the dipteran egg have been achieved by Schubiger (reviewed in
Schubiger and Newman, 1982).

By gently lowering a dull razor blade across a slightly
desiccated embryo for at least 5 min, Schubiger achieved a precocious
cytoplasmic cleavage of the embryo into "blastomeres." Subsequent
development of the "blastomeres" was identical whether the blade was
left in place or removed, indicating that the cleavage was complete
and permanent. The amount of development of the isolated
"blastomeres" varied depending upon the time at which the embryo was
precociously divided. If the embryo was cleaved during pre-blastema
stages, each "blastomere" developed into either an anterior or
posterior portion of the embryo. However, a central region of the
embryo was missing. By fusing these "blastomeres" prior to
cellularization, Schubiger et al. (1977), showed that this gap in the
mosaic structure was not due to an injury at the cleavage site but to
the premature segregation of the anterior and posterior cytoplasms.
When the cleavage was accomplished during blastema stages, the gap in
larval structure became progressively smaller until at the blastoderm
stage, the parts of the embryo equalled a whole.

These experiments clearly demonstrate the absence of a detailed
mosaic structure built into the Drosophila egg. They also suggest
that at the blastoderm stage (or the first time that the egg has
become cleaved into cells), the anterior and posterior portions of
the embryo have become autonomous and the embryo has become a mosaic
of cell types. This result has been more extensively documented by
Lohs Schardin et al. (1979a) and Underwood et al. (1980b). In the
first set of experiments, small groups of cells were killed with a
laser microbeam and the resultant larval defects analyzed. In the
second set of experiments, small numbers of cells were removed with a

micropipet and the embryos analyzed for the deficiency. In both
cases the results were remarkably consistent (Fig. 1). There was a
linear relationship between the position of damage to the blastoderm
and the segment affected. Since the number of cells on the
blastoderm is known (Turner and Mahowald, 1976), it is possible to
estimate that the primordium of each segment is comprised of a band
of cells approximately three cells wide and 40 cells long (cf.,
Lawrence and Morata, 1977). These results also illustrate that
blastoderm cells surrounding an ablation site are often unable to
regulate and replace the missing cells, suggesting that in addition
to their segmental determination, blastoderm cells may be determined
to become specific structures within larval segments.

    Blastoderm cells which contribute to adult structures have also
been shown, through use of genetic techniques, to be determined for
individual segments (Garcia-Bellido et al., 1973; Crick and Lawrence,
1975; Lawrence, 1981). It has been possible, however, to demonstrate
that those cells are not determined for specific structures within
segments. Wieschaus and Gehring (1976), observed that X-ray induced
clones of blastoderm cells can contribute to both wing and
mesothoracic leg. Furthermore, although UV microbeam ablation of
patches of blastoderm cells can produce adult cuticular defects (Lohs
Schardin et al., 1979b), normal tissue patterns can be obtained after
random, X-ray induced death of as many as 50% of the cells in an
imaginal disk (Haynie and Bryant, 1977). This result indicates that
cells within an adult compartment are capable of regenerating missing
cells. At present, it is not possible to extrapolate these results
to the blastoderm cells which are precursors to larval structures.
While agametic flies have been obtained from embryos following X-ray
doses which are thought to kill 50% of the germ cell precursors
(Wieschaus and Szabad, 1979), the fate of the blastoderm cells
producing larval structures in these embryos was not examined. It is
possible that they were not affected to the same degree as germ cells
or imaginal precursors because they undergo only a few cell
divisions. Thus, the question of whether larval epidermal lineages
are capable of regulating following random cell death remains
unanswered.

    Recently Simcox and Sang (1983) have been able to directly test
the segmental determination of blastoderm cells which contribute to
adult structures by transplanting genetically marked cells between
blastoderm embryos. They found that ectodermal precursors could
become integrated in their new sites, but in many instances still
differentiated according to their site of origin (Fig. 2). These
results convincingly demonstrate that these cells are determined for
a specific segmental quality and retain this specification even
following transplantation.

    All of the work cited so far has characterized only ectodermal
derivatives in the larva and adult. By means of individual cell

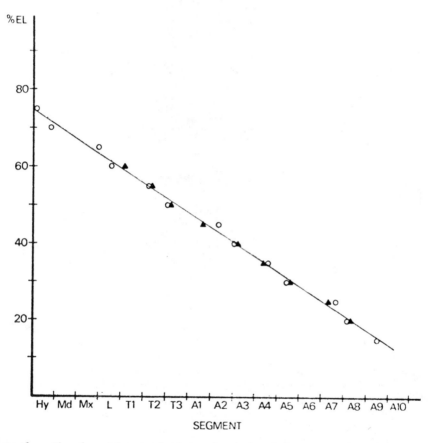

Figure 1: The locations of the embryonic defects produced by
 microbeam killing (Δ) and by ablation of cells (o) are indicated
 according to the segment affected (abscissa) and level of the
 embryo damaged (ordinate) (0% egg length is the posterior tip, and
 100% egg length is the anterior tip). The figure is adapted from
 Schubiger and Newman, 1982.

Figure 2: Figure from Simcox and Sang (1983) showing a mosaic third
          leg in a female fly derived from a blastoderm embryo which had
          received a cell transplant. The male donor was missing a first leg
          where the cells were removed and the host had first leg cells (note
          the sex combs) integrated into the third leg. The broken white
          line bounds the donor tissues. (This figure used with permission
          of Academic Press and the authors.)

transplants ventral blastoderm cells (presumptive mesodermal cells)
have been shown to be capable of differentiating into adult
endodermal cell types (reviewed by Illmensee, 1978). This clearly
indicates that these cells are not determined for formation of
mesodermal tissues. A further question concerns whether these cells
are determined for segmental identity. Lawrence and Brower (1982)
showed that mesodermal precursors from a transplanted portion of a
genetically marked wing imaginal disk could contribute to the adult
musculature throughout the thorax and abdomen of the host. These
results have been extended by the examination of X-ray induced

blastoderm clones of mesodermal precursors. Such clones were
observed to cross segmental boundaries in the adult abdomen (Lawrence
and Johnston, 1982), but not in the thorax (Lawrence, 1982). This
observation is at first paradoxical because the results of the wing
disk transplant demonstrated that the mesodermal precursors of
thoracic structures are not segmentally determined and, in addition,
that thoracic structures do not present a barrier for muscle
precursor movement. If, however, the mesodermal precursors of the
adult thorax, but not the abdomen, become associated with segmentally
determined imaginal disk precursors at the gastrula stage, then these
cell interactions might serve to constrain the mesodermal precursors
of the thorax to individual segments.

The results of Underwood, Hagel and Caulton (unpublished
results) in Mahowald's laboratory have extended these observations to
the larval mesodermal precursors. Individual ^{3}H-thymidine labeled
blastoderm cells were transplanted from one embryo to an unlabeled
host embryo and autoradiographs of the host were prepared just prior
to hatching (Table 1). In many instances the cell failed to divide,
but in ten embryos the results clearly demonstrated that the
transplanted cell was able to produce derivatives which contributed
to mesodermal structures in more than one segment or type of tissue
(Fig. 3).

Thus, it appears that at the blastoderm stage the ectodermal
precursors of both larval and adult structures are determined for
segmental identity. Although the adult epidermal precursors are
clearly not determined for specific structures within each segment,
this situation is not clear with respect to the larval epidermal
precursors. In contrast, the mesodermal and endodermal precursors of
both the larva and adult appear to be undetermined with respect to
segmental identity or restriction to mesoderm or endoderm formation.

Genetic Analysis of the Maternal Contribution to Determination

A. Maternal effect mutations can interfere with cell
determination in a global manner. During the past decade several
laboratories have performed mutagenesis screens with the goal of
isolating female sterile mutations which interfere with functions
specific to oogenesis (Rice, 1973; Gans et al., 1975; Mohler, 1977;
Engstrom et al., unpublished results). A subset of these mutations
disrupt embryonic development after cellularization (Rice and Garen,
1975; Zalokar et al., 1975). A number of mutations in this group are
now being analyzed because they affect maternal gene products which
appear to be required for normal cellular determination.

Strict maternal effect mutations which interfere with cell
determination often appear to affect the entire embryo. One of
these, dicephalic (dic), is unusual in that it also affects the
polarity of the egg (Lohs-Schardin, 1982). During normal oogenesis

Table 1. Developmental Fate of Individual Transplanted Primary
          Mesoderm Cells[a]

| | | Muscle | | Tissue | | |
Segment	body wall	inner anterior	pharynx	Around Gut[b]	Hypoderm	Nervous Tissue
Unsure of Segment	2[c]		5			
T1	1	1				
Apodeme T1/T2	2					
T2	6				2	
Apodeme T2/T3	7				1	
T3	11	2				
Apodeme T3/A1	1					
A1	2					
Mid-embryo				8		1
Posterior of Embryo				1		
Total # cells (%)	33(61%)	3(6%)	5(9)	9(17%)	3(6%)	1(2%)
# Embryos (%)	17(61%)	3(11%)	4(14%)	3(11%)	3(10%)	1(4%)

[a]Cells labeled with ^{3}H-thymidine (cf. Underwood et al., 1980a), were
   taken from the ventral midline of blastoderm/early gastrula
   staged embryos at approximately 50%EL and implanted singly into
   similar locations of host embryos. The location of labelled
   cells was analyzed at approximately 20 to 22 hr of development
   at 25°C. Numbers in the table represent actual cell counts.
[b]Cells in this group were either in muscle or fat body.
[c]The anterior-most portion of this embryo was missing, so the exact
   segmental location could not be determined.

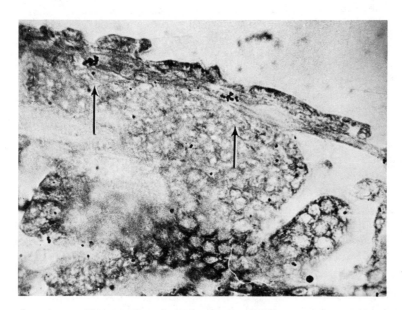

Figure 3: Autoradiograph of an embryo which received a single
mid-ventral early gastrula cell labeled with ^{3}H-thymidine. This
cell contributed progeny to segmental muscles of adjacent segments
(arrows) and to the pharyngeal musculature which is not shown in
this photograph. (Courtesy of Underwood et al., unpublished.)

the oocyte occupies the most posterior position of the 16 germ line
cells within the follicle. The remaining, more anterior 15 cells
become nurse cells. Follicle cells surrounding the germ cell cluster
appear columnar around the oocyte and squamous around the nurse
cells. Some follicle cells become border cells which move between
the nurse cells and the oocyte, and are responsible for formation of
the micropyle, an anterior egg structure.

    In contrast, the oocyte in dic follicles is situated centrally
within the follicle with nurse cells located both anteriorly and
posteriorly. Follicle cells which surround the abnormal germ cell
cluster are squamous around the nurse cells at both poles. Two sets
of border cells form, yielding eggs displaying a micropyle at each
end. The majority of embryos obtained from dic eggs develop with
anterior structures at both ends and a plane of polarity reversal
between the tips. Although the amount of anterior development is
variable among different embryos, it is similar at both ends of
individual embryos, suggesting that both ends are affected
coordinately. This mutation clearly indicates that at least some of
the information required for anterior-posterior polarity in the
embryo is dependent upon the assymetric construction of the egg and
may already be established at the time of fertilization. It is

unusual because most female sterile mutations which act early enough to affect egg shape fail to allow any embryonic development.

The recessive mutation bicaudal (Nüsslein-Volhard, 1977) also affects anterior-posterior polarity but has an effect opposite to that of dicephalic. While the morphological features of oogenesis are normal, embryos produced from bicaudal eggs display a range of mirror image duplications of the posterior end. As in embryos from dicephalic eggs the amount of posterior structure at each end of individual bicaudal embryos is generally equal, suggesting that both ends have been affected coordinately. Recently, two additional loci have been found with a bicaudal-like phenotype (Nüsslein-Volhard et al., 1982).

Maternal effect mutations also affect the global pattern of determination in the dorsal-ventral axis. The mutation dorsal (Nüsslein-Volhard, 1979) was the first mutant of this type to be characterized. Embryos from eggs produced by homozygous dl females at 25°C form a normal cellular blastoderm but fail to undergo normal gastrulation. The ventral furrow is absent and, instead of undergoing germ band elongation, the blastoderm cell layer forms a series of transverse lateral folds which continue to deepen and eventually form a string of interconnected vesicles. The cuticle which forms completely lacks lateral and ventral structures and appears to be dorsalized. Rudimentary anterior and posterior structures which form indicate that a normal anterior-posterior polarity has been maintained. This mutant also displays a dominant phenotype. Embryos from eggs produced by heterozygous females at 29°C fail to form a ventral furrow and lack mesodermal-derived tissues. In these embryos, however, ventral cuticular structures develop, suggesting that the dorsalization is less severe. Nusslein-Volhard et al. (1980) have utilized UV-microbeam ablation of blastoderm cells to determine that the fate of ventral and lateral cells in dl dominant embryos has been altered. Laser destruction of mid-ventral cells in normal embryos generally produces no cuticular defects because they represent mesodermal precursors which become internalized during ventral furrow formation. In contrast, UV killing of mid-ventral cells in dl dominant embryos produces defects in the ventral cuticule.

The manner in which these global maternal effect mutations change embryonic cell determination convincingly fits models utilizing the existence of graded concentrations of morphogens rather than localized determinants (Nüsslein-Volhard, 1979; Nüsslein-Volhard et al., 1982). The mutations are thought to act early in development to disrupt the gradients to varying degrees in either the anterior-posterior axis or the dorsal-ventral axis. It is difficult to utilize these models, however, to address the problem of how or when the blastoderm cells become determined. The dl dominant ablation studies suggest that mid-ventral cells which normally form mesoderm have become ventral hypoderm precursors. Since the evidence presented earlier indicates that the mid-ventral blastoderm cells are not determined at the time of cellularization, the interpretation of these results is actually quite difficult. For example, it is possible to imagine that rather than disrupting a global system of determination, a mutation could disrupt the ability of cells to interpret or respond to those determinative cues. Observations made in our laboratory in a developmental analysis of the maternal effect mutation gastrulation-defective (fs(1)dg), which is similar in phenotype to the dorsal recessive phenotype, yield some insights into the complexity of this problem.

The first allele at the gd locus was identified by Gans et al. (1975) and briefly described by Zalokar et al. (1975) under the name A573. Eleven additional alleles have since been isolated by Mohler (1977) and in our laboratory, and we have renamed the locus to correspond to its phenotype (Konrad et al., 1982). These alleles produce a range of phenotypes in which the most extreme is identical to the dl recessive phenotype and the most moderate appears to be similar to the dl dominant phenotype. The various gd alleles display a pattern of inter-allelic complementation either by reducing the severity of the embryonic phenotype or by hatching, which indicates that the locus is genetically complex (Konrad et al., in preparation). If we consider the variations in phenotypes to represent decreasing levels of gd locus activity, then we can hypothesize that as gd locus activity decreases, the fate of the blastoderm cells becomes increasingly dorsalized. It should be emphasized, however, that the "dorsalization" is not as direct a transformation as is seen in dicephalic or bicaudal embryos. The spinules characteristic of dorsal cuticle which form are disorganized and the process of vesiculation does not correspond to the formation of normal dorsal structures.

Although the external phenotype of embryos produced by females homozygous for extreme gd alleles is indistinguishable from that reported for dorsal embryos, we have observed differentiation of internal cells by 8-10 hr after fertilization. It is not clear how these cells become internalized but it is possible that they are either shed from the blastoderm surface or pinched off from the deep transverse folds. By 16-24 hr of development differentiated structures corresponding to endodermal and neural derivatives can be identified microscopically (Fig. 4). The presence of nervous tissue can also be detected histochemically by staining for acetylcholinesterase activity. This activity can be found throughout the string of vesicles, but is most often seen at each end. Contractile fibrils were also observed in the vesicles at a low frequency. We have obtained similar results in our examinations of dorsal recessive embryos (Konrad and Mahowald, unpublished observations).

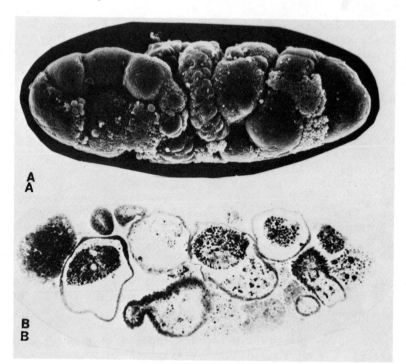

Figure 4: Phenotype of extreme gastrulation defective embryos. A
scanning electron micrograph of a gd embryo about 14 hr
post-fertilization is shown in 4A. Vesiculation has occurred and
the string of vesicles has become twisted (photograph courtesy of
F.R. Turner). A saggital section through a similar embryo about 1
hr post-fertilization is shown in 4B. The regions of differential
staining are indicative of different cell types present within the
vesicles.

In development of wild type Drosophila embryos nearly all
internal structures are derived from ventral blastoderm cells
(mesodermal and neural), or from cells at the anterior and posterior
tips (endodermal). The presence of internal structures within gd and
dl embryos does not, therefore, support a dorsalization of the fate
of all cells at blastoderm. (The major exceptions to this statement
are the nerve cell precursors which contribute to the brain of the
embryo. These have been fate mapped to the anterior dorsal region of
the blastoderm surface. While these cells could explain the presence
of acetylcholinesterase staining at one end, they cannot be the
source of neural cells at the posterior tip of these embryos.) It
appears possible that the ventral blastoderm cells in these embryos
are not determined to form dorsalized hypoderm but that they are

incapable of undergoing gastrulation. When left as part of the blastoderm surface they form cuticle which, for unknown reasons, appears to be dorsalized. When, however, they are fortuitously placed in an environment which allows them to differentiate in a more normal fashion, they can form internal structures. This hypothesis is currently being tested in our laboratory by transplanting cells from mutant embryos into wild type embryos.

A major difference between gd and dl occurs in the observable time of activity of these loci. Through use of shifts to and from the restrictive temperature we have been able to define a temperature sensitive period for one allele of gd which begins late in oogenesis and extends to 90 min post-fertilization. This early activity which occurs well before gastrulation and significant zygotic gene expression provides a reasonable explanation for the strict maternal effect and global nature of this defect. The temperature sensitivity for dl occurs as a dominant phenotype in specific genetic backgrounds and the critical period extends up to blastoderm formation (150 min) (Anderson and Nüsslein-Volhard, 1983). Injections of wild type cytoplasm into eggs from homozygous mothers can also partially rescue dl embryos up to blastoderm formation but has no effect on gd embryos (Anderson and Nüsslein-Volhard, 1982).

In summary, the gd defect does not interfere with cell formation. The defective embryos survive for a considerable period of time and adult flies which are homozygous or hemizygous for this mutation show no defect other than female sterility. This suggests that the defect is not in a common metabolic or structural gene product. The mutation does not eliminate the gene expression required for the differentiation of internal structures but rather appears to primarily affect their organization through disrupting factors required for gastrulation.

At least eight more loci have been found with phenotypes similar to gd and dl (Anderson and Nüsslein-Volhard, 1983). The appearance of the same phenotype from many loci indicates that a series of gene products must interact to produce the cellular organization requisite for normal cellular determination and cell movements. While the end points are nearly identical, dl and gd are distinguishable by their temperature sensitive period and the absence of dominance in gd. Further molecular and developmental analyses should increase our understanding of how these ten gene loci interact to produce normal dorsal-ventral polarity and gastrulation.

Maternal gene products also affect global determination events which do not involve embryonic polarity. Gene products of the daughterless (da), locus appear to be responsible for establishing dosage compensation in all of the somatic cells of the embryos and regulating the zygotic expression of the Sex-lethal locus (Sxl). In

embryos from da homozygous females, the female specific Sxl locus
appears not to be activated, causing all female offspring to die.

Another locus, extra sex comb (esc), is essential for the
establishment of segmental identity (Struhl, 1981). In the absence
of this gene product, the correct number of larval segments form but
they all resemble abdominal segment eight, suggesting that esc$^+$ may
act as a repressor of the zygotic expression of the bithorax
gene complex. Determination of the time of esc$^+$ activity using a
temperature sensitive allele has shown that esc$^+$ affects
establishment of segmental identity during a discreet period of
several hours between the end of blastoderm formation and germ band
shortening (Struhl and Brower, 1982). Since blastoderm cells of the
larval epidermal lineage are able to maintain their segmental
identity following transplantation (Simcox and Sang, 1983), this
result suggests that the genetic program of differentiation has been
initiated but not stabilized at blastoderm. This stabilization, at
least with respect to esc$^+$ expression, may occur progressively
following blastoderm formation. The Polycomb (Pc) locus is defined
by dominant mutations which produce a phenotype similar to esc. Pc
also displays a maternal effect and appears to interact with both the
bithorax and antennapedia gene complexes (Denell and Frederick,
1983). In contrast to esc, Pc appears to be required throughout
development to maintain segmental identity. When homozygous Pc
clones were induced during larval development, the eye-antennal and
thoracic disks were observed to differentiate analia-like tissue
(Struhl, 1981), perhaps corresponding to the eighth abdominal segment
of the adult (Struhl and Brower, 1982).

B. Localized defects produced by maternal effect mutations.
Not all maternal effect mutations affecting embryonic development
produce global alterations. A detailed description of fs(1)1502 has
been made by Komorowska (1980), and of a new allele of this locus by
Turner and Stephenson of our laboratory (unpublished observations).
Embryos derived from homozygous mothers lack the clypeolabrum or
anterior-most head segment and the resultant embryos have defective
mouthparts and rarely hatch.

We have also found two loci on the X-chromosome
(fs(1)E1;fs(1)E2) in which the embryos from mutant flies fail to
develop structures which derive from the anterior and posterior tips
of the blastoderm. The phenotype is similar for both loci. The
first visible defect is a deficiency at the posterior tip below the
pole cells (Fig. 5A). Subsequently, all of the structures that
should form from the posterior 20% of the blastoderm (for fate maps,
see Poulson, 1950, and Underwood et al., 1980b) are missing. At the
anterior tip the mouthparts resemble fs(1)1502 so that there may be
anterior defects in E1 and E2 similar to those associated with 1502.
Thus, in these two mutations the fate map of the blastoderm appears
truncated at both ends with the development of the remainder of the

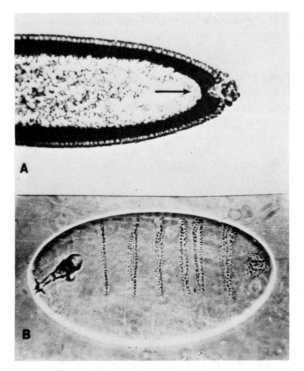

Figure 5: Phenotype of fs(1) ± El embryos. A saggital section at
    the blastoderm stage (5A) shows localized degeneration of the
    blastoderm below the pole cells (arrow). Subsequently, no midgut
    or proctodeal invagination occurs during gastrulation and following
    segmentation (5B); structures beyond abdominal segment 7 are
    missing from the posterior end (right) of the embryo. The mouth
    parts at the anterior end (left) of the embryo are also abnormal.

embryo relatively normal (Fig. 5B). These mutants have a phenotype
similar to the second chromosomal mutant fs(2)torso briefly described
by Nusslein-Volhard et al. (1982). The interesting observation
concerning these last mutations is that although they are strict,
maternal effect embryonic lethals, they apparently affect limited
portions of the blastoderm fate map, in contrast to the more global
defects produced by bicaudal or dorsal.

    C. Some maternal effect mutations can be rescued by the zygotic
genome. We have recently begun the investigation of a class of
maternal effect mutations which display a rescuable phenotype. In
contrast to the strict maternal effect lethal mutations considered
above, the embryonic lethality of this group is either partially or
completely alleviated by the introduction of the wild-type allele
with the sperm. While this result clearly suggests that these loci

are active during embryogenesis, this embryonic gene activity is not
needed since homozygous mutant embryos from heterozygous mothers can
survive.

A number of female sterility mutations of this class have been
shown to produce metabolic deficiencies. For example, rudimentary is
responsible for three enzymes of pyrimidine biosynthesis (Norby,
1973), and deep orange (Counce, 1956a) and cinammon (Baker, 1973) are
involved in the production of the eye pigments. Because of their
metabolic involvement, there have been no extensive developmental
analyses of this class of mutants.

Other examples of this group of mutants, however, have a much
more defined lethal phenotype caused by defects in the proper
determination of specific groups of cells at the blastoderm stage.
While the lesion in these loci is not known, it is less likely that a
general metabolic deficiency can produce so specific a developmental
defect.

Both almondex (amx) (Shannon, 1972, 1973; Jimenex and
Campos-Ortega, 1982) and fs(1)R1 (Romans et al., 1976; LaBonne et
al., unpublished observations) show a maternal effect phenotype
resembling the nervous system hypertrophy characteristic of Notch
(Poulson, 1940; Wright, 1970) and other neurogenic mutants (Lehmann
et al., 1981). In these mutants the whole ventral ectoderm develops
according to the neural lineage rather than only a portion becoming
neural. While this change in determination is fully penetrant in amx
and fs(1)R1 when homozygous females are mated with mutant males,
partial rescue of the phenotype is produced by introducing the wild
type allele with the sperm. Recent results from our laboratory
(LaBonne and Mahowald, unpublished observations) indicate that both
amx and fs(1)R1 are also partially rescued by microinjection of wild
type cytoplasm into preblastoderm embryos.

In the maternal effect mutant fused (Counce, 1956b; Mahowald,
unpublished observations), the ectodermal cells of the posterior
compartment of each abdominal segment differentiate as a mirror image
of the anterior compartment, thus producing a "gooseberry" pattern
for the larval ventral cuticle (cf. Nusslein-Volhard and Wieschaus,
1980). The developmental aberration, while formed with full
penetrance when homozygous females are mated with mutant males, is
partially corrected when the wild type allele is introduced with the
sperm. The previously mentioned esc locus also fits this
classification because the maternal defect can be rescued to a large
extent by the introduction of two copies of the esc^+ gene by the
sperm (Struhl, 1981).

Finally, we have discovered fs(1)E3 (Engstrom et al.,
unpublished observations) in which embryos produced from mutant
females have a limited transformation of a small region of the

posterior blastoderm. This region is transformed into ill-defined new structures which protrude from the posterior ventral surface of the embryo during post-gastrulation stages of development. In contrast to the previously mentioned loci, fs(1)E3 is completely rescued by the wild type allele introduced with the sperm.

The fact that the gene products of these loci can be introduced into embryos during oogenesis, via zygotic gene activity, or even by microinjection into the preblastoderm embryo indicates that they are not part of the primary ooplasmic components for establishing embryonic axes. They must, however, be present for the proper determination and normal development of specific sets of blastoderm cells. A reasonable model suggests that these gene products must interact with other components of the early embryo which previously establish the basic polarities of the embryo.

CONCLUSION

The maternal effect mutations studied to date in Drosophila indicate that determinative information in the egg is responsible for early, global developmental events such as establishment of egg polarity and the initiation of dosage compensation. Maternal effect mutations also affect apparently later developmental events such as the establishment of segmental identity and compartmental polarity as well as more localized differentiations. A large number of zygotic mutations have been found to affect segment number and polarity (Nusslein-Volhard and Wieschaus, 1980). Since the number of segments and compartments appears to be progressively established between fertilization and blastoderm formation, interactions between maternal information and the zygotic genome are occurring during this time.

Models suggesting how zygotic gene expression could elaborate upon and refine maternally established polarity have been proposed (Capdevilla and Garcia-Bellido, 1981; Kaufman and Wakimoto, 1982). In general, these models hypothesize gradients of maternal information similar to those mentioned earlier with respect to the interpretation of dorsal, bicaudal and dicephalic.

While it may be the simplest model to have monotonic gradients in both the anterior-posterior and dorsal-ventral planes responsible for every range of differentiation, it is more reasonable to separate the model for segment determination from endodermal and mesodermal derivatives. Both of these latter cell types require specific cell movements and interactions prior to recognizable differentiations. Our examination of the gd locus suggests that aspects of these interactions may also be dependent upon maternal information.

Possibly the most exciting result arising from recent screens of maternal effect lethal mutations is the discovery that multiple loci produce similar phenotypes (also cf. Anderson and Nusslein-Volhard, 1983). For example, there are now ten dorsal-like, three bicaudal-like, and at least three torso-like loci. It is clear that the establishment of embryonic polarity requires the interaction of many gene products. Only two loci affecting dorsal-ventral polarity have been studied to any extent and they differ in a number of important developmental aspects. Thus, dorsal displays a dominant phenotype which is lacking in gastrulation-defective alleles. In addition, gd embryos have a temperature sensitive period which is completed prior to the blastoderm stage while the temperature sensitive period of dorsal embryos continues into the blastoderm stage. Further analysis of these sets of maternal effect loci should provide a new understanding of the role of maternal information in establishing embryonic patterns of development.

ACKNOWLEDGEMENTS

We gratefully acknowledge the efforts and expertise of Joan Caulton and Rudy Turner in the initial characterization of many of the mutants discussed in this paper, and Eileen Underwood, Susan Hagel and Joan Caulton in the mesodermal cell transplant studies. In addition, we would like to especially thank Lee Engstrom, Norbert Perrimon, Tom Goralski and Steve LaBonne for many stimulating discussions of the genetics of early development. This work was supported by NIH HD 17607-02 to A.P.M.

REFERENCES

Anderson, D. (1966). The comparative embryology of the Diptera. Ann. Rev. Ent. 11:23-46.

Anderson, K. and Nusslein-Volhard, C. (1982). Genetic control of dorsal-ventral polarity in the Drosophila embryo. Biology of the Cell 45:400.

Anderson, K.V. and Nusslein-Volhard, C. (1983). Genetic analysis of dorsal-ventral embryonic pattern in Drosophila. In "Primers in Developmental Biology," G. Malacinski and S. Bryant (eds.). MacMillan Press, in press.

Baker, B. (1973). The maternal and zygotic control of development by cinnamon, a new mutant in Drosophila melanogaster. Develop. Biol. 33:429-440.

Capdevila, M.P. and Garcia-Bellido, A. (1981). Genes involved in the activation of the bithorax complex of Drosophila. Wilhelm Roux' Arch. 190:339-350.

Crick, F.H.C. and Lawrence, P.A. (1975). Compartments and polyclones in insect development. Science 189:340-347.

Counce, S.J. (1956a). Studies on female-sterility genes in Drosophila melanogaster: I. The effects of the gene deep orange on embryonic development. Z. VererbLehre 87:443-461.

Counce, S.J. (1956b). Studies on female-sterility genes in Drosophila melanogaster: II. The effects of the gene fused on embryonic development. Z. VererbLehre 87:462-481.

Davidson, E.H. (1976). Gene Activity in Early Development. Academic Press, New York.

Denell, R.E. and Frederick, R.D. (1983). Homeosis in Drosophila: A description of the Polycomb lethal syndrome. Dev. Biol., in press.

Gans, M., Audit, C., Masson, M. (1975). Isolation and characterization of sex-linked female sterile mutants in Drosophila melanogaster. Genetics 81:683-704.

Garcia-Bellido, A., Lawrence, P.A., and Morata, G. (1973). Developmental compartmentalization of the wing disk of Drosophila. Nature New Biol. 245:251-253.

Haynie, J.L. and Bryant, P.J. (1977). The effects of X-rays on the proliferation dynamics of cells in the imaginal wing disk of Drosophila melanogaster. Wilhelm Roux' Arch. 183:85-100.

Illmensee, K. (1978). Drosophila chimeras and the problem of determination. In "Genetic Mosaics and Cell Differentiation," W.J. Gehring (ed.), pp. 51-69. Springer-Verlag, New York.

Jimenez, G. and Campos-Ortega, J.A. (1982). Maternal effects of zygotic mutants affecting early neurogenesis in Drosophila. Wilhelm Roux' Arch. 191191-201.

Kaufman, T.C. and Wakimoto, B.T. (1982). Genes that control high-level developmental switches. In "Evolution and Development," J.T. Bonner (ed.), pp. 189-205, Dahlem Konferenzen. Springer-Verlag, New York.

Komorowska, B. (1980). L'effet d'une mutation de sterilite femelle sur la cephalogenese de Drosophila melanogaster pendant la vie embryonnaire. Thesis, University of P. and M. Curie.

Konrad, K.D., Goralski, T.J., Turner, F.R., and Mahowald, A.P. (1982). Maternal effect mutation affecting gastrulation. J. Cell Biol. 95:159a.

Lawrence, P.A. (1981). The cellular basis of segmentation in insects. Cell 26:3-10.

Lawrence, P.A. (1982). Cell lineage of the thoracic muscles of Drosophila. Cell 29:493-503.

Lawrence, P.A. and Brower, D.L. (1982). Myoblasts from Drosophila wing disks can contribute to developing muscles throughout the fly. Nature 295:55-57.

Lawrence, P.A. and Johnston, P. (1982). Cell lineage of the Drosophila abdomen: The epidermis, oonocytes, and ventral muscles. J. Embryol. Exp. Morph. 72:197-208.

Lawrence, P.A. and Morata, G. (1977). The early development of mesothoracic compartments in Drosophila. Dev. Biol. 56:40-51.

Lehmann, R., Dietrich, U., Jimenez, F., and Campos-Ortega, J.A. (1981). Mutations of early neurogenesis in Drosophila. Wilhelm Roux' Arch. 190:226-229.

Lohs-Schardin, M. (1982). Dicephalic--A Drosophila mutant affecting polarity in follicle organization and embryonic patterning. Wilhelm Roux' Arch. 191:28-36.

Lohs-Schardin, M., Cremer, C., and Nusslein-Volhard, C. (1979a). A fate map for the larval epidermis of Drosophila melanogaster: Localized cuticle defects following irradiation of the blastoderm with an ultraviolet laser microbeam. Develop. Biol. 73:239-255.

Lohs-Schardin, M., Sander, K., Cremer, C., Cremer, T., and Zorn, C. (1979b). Localized ultraviolet laser microbeam irradiation of early Drosophila embryos: Fate maps based on location and frequency of adult defects. Develop. Biol. 68:533-545.

Mahowald, A.P. (1977). The germ plasm of Drosophila: An experimental system for the analysis of determination. Amer. Zool. 17:551-563.

Mahowald, A.P. and Boswell, R.E. (1983). Germ plasm and germ cell development in invertebrates. In "Current Problems in Germ Cell Differentiation," McLaren and Wylie (eds.). Cambridge Press, in press.

Mohler, J.D. (1977). Developmental genetics of the Drosophila egg: I. Identification of 59 sex-linked cistrons with maternal effects on embryonic development. Genetics 85:259-272.

Norby, S. (1973). The biochemical genetics of rudimentary mutants
    of Drosophila melanogaster. Hereditas. 73:11-16.

Nusslein-Volhard, C. (1977). Genetic analysis of pattern formation
    in the embryo of Drosophila melanogaster. Characterization of
    the maternal-effect mutant bicaudal. Wilhelm Roux' Arch.
    183:249-268.

Nusslein-Volhard, C. (1979). Maternal effect mutations that alter
    the spatial coordinates of the embryo of Drosophila
    melanogaster. In "Spatial Determinants of Development," S.
    Subtelny and I. Konigberg (eds.), pp. 185-211. Academic Press,
    New York.

Nusslein-Volhard, C., Lohs-Schardin, M., Sander, K., and Cremer, C.
    (1980). A dorso-ventral shift of embryonic primordia in a new
    maternal-effect mutant of Drosophila. Nature 283:474-476.

Nusslein-Volhard, C. and Wieschaus, E. (1980). Mutations affecting
    segment number and polarity in Drosophila. Nature 287:795-801.

Nusslein-Volhard, C., Wieschaus, E., and Jurgens, G. (1982).
    Segmentation in Drosophila, a genetic analysis (Segmentierung in
    Drosophila, Eine genetische Analyse). In "Veshandlungen des
    deutschen Zoologischen Gesellschaft." Guster Fischer Verlag, in
    press.

Poulson, D.F. (1940). The effects of certain X-chromosome
    deficiencies on the development of Drosophila melanogaster. J.
    Exp. Zool. 83:271-325.

Poulson, D.F. (1950). Histogenesis, organogenesis, and
    differentiation in the embryo of Drosophila melanogaster Meigen.
    In "Biology of Drosophila," M. Demerec (ed.), pp. 168-274.
    Wiley, New York.

Rice, T.B. (1973). Isolation and characterization of
    maternal-effect mutants: an approach to the study of early
    determination in Drosophila melanogaster. Ph.D. Dissertation,
    Yale University, New Haven, 147 pp.

Rice, T.B. and Garen, A. (1975). Localized defects of blastoderm
    formation in maternal effect mutants of Drosophila. Develop.
    Biol. 43:277-286.

Romans, P., Hodgetts, R.B., and Nash, D. (1976). Maternally
    influenced embryonic lethality: Allele specific genetic rescue
    at female fertility locus in Drosophila melanogaster. Can. J.
    Genet. Cytol. 18:773-781.

Shannon, M.P. (1972). Characterization of the female-sterile mutant almondex of Drosophila melanogaster. Genetica 43:244-256.

Shannon, M.P. (1973). The development of eggs produced by the female-sterile mutant almondex of Drosophila melanogaster. J. Exp. Zool. 183:383-400.

Schubiger, G., Moseley, R.C., and Wood, W.J. (1977). Interaction of different egg parts in determination of various body regions in Drosophila melanogaster. PNAS 74:2050-2053.

Schubiger, G. and Newman Jr., S.M. (1982). Determination in Drosophila embryos. Amer. Zool. 22:47-55.

Simcox, A.A. and Sang, J.H. (1983). When does determination occur in Drosophila embryos? Develop. Biol., in press.

Struhl, G. (1981). A gene product required for correct initiation of segmental determination in Drosophila. Nature 293:36-41.

Struhl, B. and Brower, D. (1982). Early role of the esc$^+$ gene product in the determination of segments in Drosophila. Cell 31:285-292.

Turner, F.R. and Mahowald, A.P. (1976). Scanning electron microscopy of Drosophila embryogenesis: I. The structure of the egg envelopes and the formation of the cellular blastoderm. Devel. Biol. 50:95-108.

Underwood, E.M., Caulton, J.H., Allis, C.D., and Mahowald, A.P. (1980a). Developmental fate of pole cells in Drosophila melanogaster. Devel. Biol. 77:303-314.

Underwood, E.M., Turner, F.R., and Mahowald, A.P. (1980b). Analysis of cell movements and fate mapping during early embryogenesis in Drosophila melanogaster. Dev. Biol. 74:286-301.

Wieschaus, E. and Gehring, W. (1976). Clonal analysis of primordial disk cells in the early embryo of Drosophila melanogaster. Devel. Biol. 50:249-263.

Wieschaus, E. and Szabad, J. (1979). The development and function of the female germ line in Drosophila melanogaster: A cell lineage study. Dev. Biol. 68:29-46.

Wright, T.R.F. (1970). The genetics of embryogenesis in Drosophila. Adv. in Genetics 15:262-395.

Zalokar, M., Audit, C., and Erk, I. (1975). Developmental defects of female-sterile mutants of Drosophila melanogaster. Devel. Biol. 47:419-432.

HOMOEOTIC GENES AND THE SPECIFICATION OF SEGMENTAL IDENTITY

IN THE EMBRYO AND ADULT THORAX OF DROSOPHILA MELANOGASTER

Thomas C. Kaufman and Michael K. Abbott[*]

ABSTRACT

The homoeotic loci Sex-Combs Reduced (Scr), Antennapedia (Antp), and Ultrabithorax (Ubx), are required early in development to specify the thoracic segmental identities of the embryo. However, each locus also has a later function(s): to maintain specific aspects of thoracic segmental identity through metamorphosis. $Antp^+$ primarily specifies mesothoracic (T_2) identity while the embryonic prothoracic (T_1) and metathoracic (T_3) identities are dependent, respectively, upon Scr^+ and Ubx^+ and interactions between these loci and $Antp^+$. Since the loss of Scr^+ and/or Ubx^+ in combination with the loss of $Antp^+$ function results in a transformation of one or more of the thoracic segments to a T_1-gnathal identity, $Antp^+$ function is at least partially epistatic to that of either Scr^+ or Ubx^+ during the establishment of thoracic segmental identities. Somewhat later in development, Scr^+ and Ubx^+ are required to maintain the identity of specific regions of either the pro- or metathorax, respectively. By contrast, $Antp^+$ functions to maintain the identities of certain regions in all three segments. Loss of this function allows portions of the ventral region of each thoracic segment to become antenna-like. In addition, a loss of identity of a region of the anterior, dorsal-lateral mesothorax also occurs. Taken together, these results indicate that Antp is a complex locus with a central role in the establishment and maintenance of thoracic segmental identities. Differences among the thoracic segments arise through the interaction of Scr and Ubx with Antp.

[*]From the Program in Genetics and Molecular, Cellular, and Developmental Biology, Department of Biology, Indiana University, Bloomington, IN 47405.

189

INTRODUCTION

    While recent investigations have revealed much about gene
structure and regulation, we still understand very little of the
exact manner in which genes function to transform a zygote into the
multitude of differentiated cells found in the adult organism. How
is this genetic control of development integrated? By what
mechanism(s) does this information control the behavior of groups of
cells to produce the coordinated, three-dimensional changes which
occur during morphogenesis? One approach toward unravelling the
solutions to these difficult questions is through the analysis of the
manner in which gene mutations affect developmental processes. That
this approach is useful is illustrated by recent investigations into
the abnormalities in the early events of mouse embryogenesis caused
by mutations in the loci of the T-complex (Bennett, 1975) or
mutations which alter the normal cell lineage patterns during the
larval development of C. elegans (Chalfie et al., 1981). In our
laboratory we have chosen to examine, by mutational analysis, genes
which control the basic segmentation pattern of Drosophila
melanogaster.

    Our investigations have focused upon the genetic and
developmental attributes of a group of homoeotic mutations.
Mutations of this nature cause or allow a transformation of the
identity of one segment or metamere into that of another. The
segment which is mimicked develops in an entirely normal manner while
the transformed segment follows a pattern of development
characteristic of the segment into which it has been transformed.
Since there is no disruption--only a simple reordering of the normal
ontogenetic process--homoeotic genes have been viewed as ontogenetic
switches that regulate the normal temporal and/or spatial expression
of the genetic information necessary for proper development
(Kauffman, 1975; Kauffman, 1977; Kauffman et al., 1978). Therefore,
the study of homoeotic loci may permit an understanding of how a
single locus may regulate the expression of several other genes in a
programmed fashion.

    Homoeotic loci are found throughout the Drosophila genome
(Ouweneel, 1976). However, there are two distinct clusters of such
loci in the right arm of the third chromosome. The more distal of
these is called the Bithorax Complex (BX-C) and has been the subject
of investigations by several laboratories--most notably that of E.B.
Lewis (Lewis, 1963, 1964, 1978, 1981a, 1981b; Garcia-Bellido, 1977;
Garcia-Bellido and Ripoll, 1978). Mutations in the BX-C result in
transformations restricted to the posterior segments of the
Drosophila embryo and adult. The proximal cluster is called the
Antennapedia Complex (ANT-C) and has been investigated in our
laboratory and others (Denell, 1973; Duncan and Kaufman, 1975;
Kaufman, 1978; Kaufman et al., 1980; Lewis et al., 1980a; Lewis et
al., 1980b; Denell et al., 1981; Wakimoto and Kaufman, 1981; Kaufman

and Wakimoto, 1982; Struhl, 1981, 1982). The chromosomal location of the ANT-C and the loci which comprise it, as well as those in the surrounding chromosomal environs, are all shown in Fig. 1. Unlike the BX-C, the members of the ANT-C cause or allow transformations among segments of the anterior end of the embryo and adult. The distribution of those regions of gene action within the embryo is presented in Fig. 2. The various loci of the complex affect the thoracic, mouthpart, and head segments and, with the exception of the fushi tarazu (ftz) locus (Wakimoto and Kaufman, 1981), do not exhibit any alterations of the abdomen. The two complexes (BX-C and ANT-C) are not entirely exclusive in their domains of activity which overlap in the region of the thorax. It is the major loci of the two complexes which specify the identity of the three segments forming this region of the animal which will be the major focus of this paper. These genes exemplify the kinds of transformations observed in homoeosis and can serve as a model system for this type of genic regulation of ontogeny. Moreover, this particular region of the animal and these genes demonstrate the hierarchial nature of the process responsible for the specification of segmental identities.

The Ultrabithorax Locus

On the third chromosome the most proximal of the sites within the BX-C is the Ultrabithorax (Ubx) locus and the various recessive mutations which fail to complement the dominant Ubx lesions. Flies heterozygous for Ubx (i.e., Ubx/+) have an enlarged haltere (balancer organ). This organ is found on the dorsal aspect of the adult metathorax and, in evolutionary terms, represents a vestigial wing. Homozygous (Ubx/Ubx) individuals die in the larval portion of the life cycle. Observations on the morphology of these animals has shown that the larval third thoracic and first abdominal segments are transformed into the likeness of the second thoracic segment, resulting in an animal with three tandemly arrayed second thoraces. This same type of transformation can be observed in adult flies by the use of the above mentioned recessive mutations. The bithorax (bx) and postbithorax (pbx) mutations transform the anterior and posterior portions, respectively, of the adult metathorax into the likeness of the mesothorax. This results in a transformation of the third leg into a second leg and the small haltere into a wing. Moreover, the normally small band of cuticle which makes up the dorsal metathorax (metanotum) is enlarged and resembles the dorsal mesothorax (mesonotum). These phenotypes are observed in either bx pbx/bx pbx or bx pbx/Ubx individuals. A third recessive mutation, bithoraxoid (bxd), causes a transformation of the adult first abdominal segment into one resembling a thoracic segment. Like bx and pbx, this can be observed in either homozygous bxd or bxd/Ubx individuals. In triply mutant bx pbx bxd individuals the adult phenotype resembles that of the Ubx lethal individuals described above. That is, the third thoracic (T_3) and first abdominal (A_1) segments resemble the second thoracic (T_2) segment. Therefore, it is

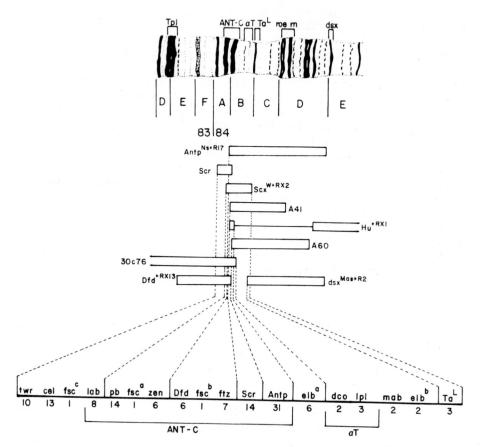

Figure 1: Drawing of the segment of the right arm of the third
chromosome containing loci of the Antennapedia Complex (ANT-C).
Open bars below the chromosome show cytological extent of a group
of deletions which encroach on and delete all or portions of the
ANT-C. The dark line below the deletions represents the genetic
map of loci within and adjacent to the ANT-C. Brackets below the
genetic map indicate the location of proposed members of the ANT-C
and those mutations which apparently affect α tubulin (αT).
Numbers below the line show the number of alleles at each locus.
Dotted lines indicate positions of various deletion end points
relative to the genetic map. Tpl = Triplo-lethal; Ta^L = Thickened
arista lethal; roe = roughened eye; rn = rotund; dsx = doublesex;
twr = twisted bristles roughened eyes; cel = cell lethal; fsc^c =
female sterile of Cain-c; lab = labial; pb = proboscipedia; fsc$_a$ =
female sterile of Cain-a; zen = zerknullt; Dfd = Deformed; fsc^b =
female sterile of Cain-b; ftz = fushi tarazu; Scr = Sex Combs
Reduced; Antp = Antennapedia; elb^a = embryo-larval boundry
lethal-a; dco = discs overgrown; lpl = larval pupal lethal; mab =
malformed abdomen; elb^b = embryo-larval boundry lethal-b.

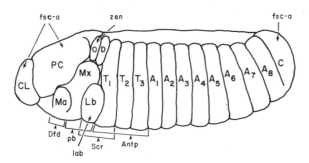

Figure 2: Drawing of the lateral aspect of an 8-hr-old <u>Drosophila</u>
embryo. At this point in development all of the segments of the
animal are clearly evident on the surface of the animal. The head
is made up of the Clypeolabral (CL), Procephalic (PC), Optic (O),
Mandibular (Ma), Maxillary (Mx), and Labial (Lb) lobes as well as
the Dorsal ridge (D). There are three thoracic (T_1, T_2, T_3), eight
abdominal (A_1 - A_8), and a terminal Caudal (C) segment which
comprise the trunk. The arrows and brackets indicate those
segments in which the various members of the ANT-C are active
either in the specification of segment identity or in the formation
of the segment <u>per se</u>. Gene abbreviations as in Fig. 1.

possible to conclude that <u>Ubx</u> and the three recessive mutations it
fails to complement are necessary for the specification of segmental
identity in T_3 and A_1 and, in the absence of their activity, a T_2
pattern is elaborated.

Since the segment identities of both the larva and adult are
affected by <u>Ubx</u>, we may conclude that this gene is normally active
early in development. This has been verified by two different
methods. The first of these involves the production of phenocopies
of the bithorax mutant phenotypes. Non-heritable somatic copies of
<u>bx</u> and <u>pbx</u> have been created by either ether or heat treatment of
early embryos (Capdevila and Garcia-Bellido, 1978). The efficacy of
the phenocopying agent is highly dependent on the state of the BX-C
(i.e., animals carrying one dose of the complex are more sensitive
than are homozygous normal or animals carrying an extra copy of the
BX-C). Therefore, it would appear that the phenocopying event is
taking place through an interaction with the complex itself or its
product(s), thereby mimicking a mutational inactivation of the <u>bx</u> or
<u>pbx</u> functions. The stage in development which is most sensitive to
the phenocopy-producing agent is the cellular blastoderm, about 2.5
hr post-fertilization. A variety of experimental results have shown
that this is the point in <u>Drosophila</u> ontogeny where the basic body
plan is determined and a majority of cellular fates are specified
(Schubiger and Wood, 1977; Schubiger and Newman, 1982). Therefore,

it is reasonable to conclude that Ubx functions in these early determinative events to specify segment identity in T_3 and A_1.

The second technique used to define the role of Ubx^+ in segment specification is x-ray induced somatic recombination (Becker, 1975). Flies heterozygous for any mutation and a wild-type allele of that locus can be irradiated at different points during development, and among the population of cells making up the animal at the time of the irradiation, infrequent exchange events will be induced between the mutant and wild-type bearing homologues. Subsequent cell division in some cases will result in the segregation of two genetically different types of daughter cells. One will be homozygous mutant, the other wild-type. These two daughters have the potential to proliferate and form a population of clonally related offspring. If the chromosome of interest is also marked with independent mutations which alter the color of the adult integument or the morphology of the bristles and hairs, the two populations of cells can be detected on the surface of an adult fly as a patch of cuticle differing in morphology from the surrounding tissue. Just such a technique has been used with Ubx and its associated recessive mutations. These studies have revealed that clones of cells homozygous for any Ubx mutations autonomously express the mutant phenotypes described above for the homozygous recessive bearing adults (Garcia-Bellido and Ripoll, 1978). This is true for clones created from just after cellular blastoderm all the way through to the pupal stage. Therefore, Ubx^+ function is not only necessary for the early specification of segment identity, it is also necessary for the maintenance of that specified state until differentiation of the adult takes place during metamorphosis of the pupa.

An additional aspect of Ubx function has recently come to light through the analysis of clones created just before or at cellular blastoderm (Morata and Kerridge, 1981; Kerridge and Morata, 1982; Minana and Garcia-Bellido, 1982). Such clones show a somewhat different transformation than those made subsequent to blastoderm formation. The anterior half of the metathoracic leg is transformed to mesothorax just as before; however, both the posterior half of the mesothoracic and metathoracic legs are transformed to prothoracic legs. The dorsal portions of these animals show the same $T_3 \rightarrow T_2$ transformations observed before. Therefore, it would appear that Ubx^+ functions somewhat differently prior to blastoderm formation, at least in the ventral portions of the animal, than it does after this point. These early requirements for Ubx^+ action are indicative of the early versus the late temporal component of Ubx expression. As will be seen, this theme will reappear in our consideration of the other two thorax-specifying loci.

The Sex Combs Reduced (Scr) Locus

The Scr locus is a member of the ANT-C (Fig. 1), and mutations

of this locus affect the labial (Lb) and prothoracic (T_1) segments
(Fig. 2). A majority of the mutant alleles are dominant and result
in a reduction in the number of sex comb teeth on the prothoracic leg
of adult male flies (from 8-12 to 4-5). This same phenotype is
observed in flies heterozygous for a deletion of the locus.
Therefore, the sex comb reduction results from an inactivation of the
locus which apparently reduces the effective gene dosage below some
threshold level. Most Scr mutant alleles are homozygous lethal and
cause the death of the embryo just prior to hatching (about 24 hr
post-fertilization) (Wakimoto and Kaufman, 1980). Two alleles,
however, are semi-lethal and a few individuals survive to become
adults. These animals show a nearly complete loss of sex comb teeth
and a transformation of their labial palps into maxillary palps (Fig.
3). We have recently recovered two temperature-sensitive alleles of
Scr, both of which survive to adulthood and show a similar phenotype.
In addition, the first leg of these animals is completely transformed
into a second leg (Otteson and Kaufman, unpublished). The
temperature-sensitive period for this transformation occurs during
the late third larval instar, just prior to pupation.

The morphology of lethal Scr individuals is shown in Fig. 4.
The morphology of the prothoracic segment resembles that of the
second thorax, and the labial segment of the embryo bears a set of
maxillary sense organs (Wakimoto and Kaufman 1980; Kaufman
unpublished). Therefore, the transformations observed in the adult,
$T_1 \rightarrow T_2$ and Lb $\rightarrow$ Mx, are also found in the embryo and larva.

Somatic recombination analyses with Scr lesions (Wakimoto, 1981;
Struhl, 1982) have shown that Scr^+, like Ubx^+, is necessary
throughout the larval stages for the maintenance of the T_1 and Lb
determined state. This result is also consistant with the late
larval temperature-sensitive period found in this laboratory. The
fact that the pattern of segmental identity is altered in the embryo
and larva is indicative of the fact that Scr most likely functions in
the early specification of segment identity in a manner similar to
Ubx.

In light of the fact that Scr and Ubx transform T_1 and T_3,
respectively, into T_2, we wondered what would be the result of a
simultaneous transformation of these segments. To this end we
created a double mutant ($Scr^- Ubx^-$) bearing chromosome. Animals
homozygous for this chromosome die at the end of embryogenesis and
have tandem, Mx-like segments followed by three T_2-like segments.
Therefore, with respect to the apparent embryonic phenotype, the two
mutations are simply additive in their effects (see Fig. 15 for a
summary). Somatic recombination performed using a chromosome similar
to our double mutant has revealed, however, that the effect is not
strictly additive. Genotypically $Scr^- Ubx^-$ clones apparently induced
at cellular blastoderm yield anterior and posterior portions of the
meso- and metathoracic legs which are entirely mesothoracic in

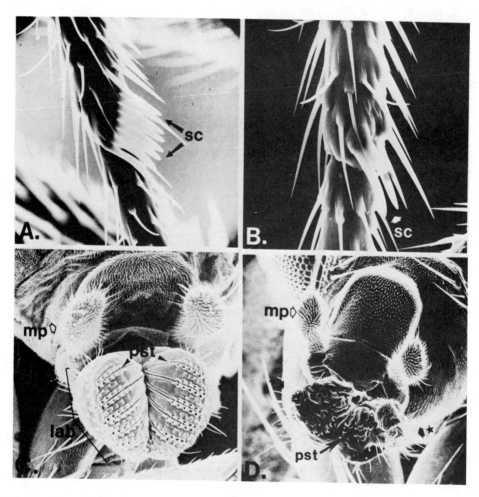

<u>Figure</u> 3: Scanning electron micrographs of the prothoracic legs (A
   and B) and labial palps (C and D) of normal (A and C) and <u>Scr</u> (B
   and D) animals. The animals are heterozygous for a semi-lethal
   allele of the locus and a deletion of the entire ANT-C [<u>Df</u>(3R)<u>Scr</u>].
   Note the marked reduction in the number of sex comb teeth (Sc) in
   B. Other aspects of the chetotaxy of the first leg indicate a
   transformation of $T_1 \rightarrow T_2$. The solid arrows and stars indicate the
   region of the labial palps (lab) which are transformed to resemble
   maxillary palps (mp). The transformation also results in a
   reduction in number and derangement of the psuedotrachea (pst).

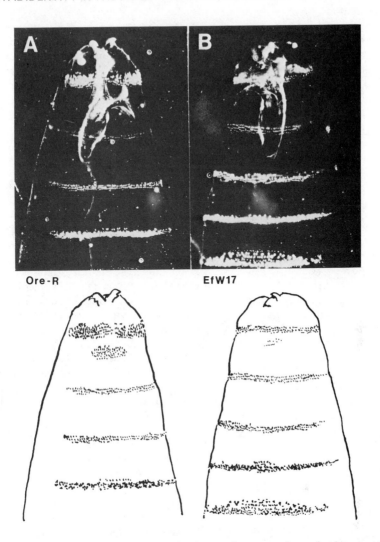

Figure 4: Darkfield illuminated photomicrographs of the ventral cuticle of late embryos of a normal (A) and an Scr mutant (B) individual. Below the micrographs are tracings showing the morphology of the denticle belts at the leading edge of each of the three thoracic segments. Anterior is at the top of each panel. Note the reduction in size and number of the denticles of the first or T_1 row in the Scr individual relative to normal. Also note that the head region anterior to the T_1 row is broader in the Scr animal.

character (Struhl, 1982). The prior result giving prothoracic structures in the meso- and metathoracic legs when only Ubx is deficient is not observed. Therefore, we may conclude that early in development (i.e., cellular blastoderm) Ubx$^+$ functions in the posterior portions of the ventral T_2 and T_3 to modify or repress the action of $\underline{Scr}^+$. This modification, once accomplished, need not be maintained since $\underline{Ubx}^-$ clones created subsequent to the event do not reveal $\underline{Scr}^+$ action. It should be emphasized, however, that $\underline{Scr}^+$ activity in T_2 and T_3 is not normal ($\underline{Scr}^-$ cells develop normally in these two segments) but only occurs in the absence of $\underline{Ubx}^+$ activity.

The Antennapedia (Antp) Locus

The $\underline{Antp}$ locus is perhaps the most complex of the loci we are considering. The genetic lesion for which the locus was named is dominant and causes a transformation of the antenna to mesothoracic leg (An → T_2) (Fig. 5). A similar type of dominant transformation is caused by the $\underline{Nasobemia}$ ($\underline{Ns}$) mutation (Gehring, 1966); the major difference residing in the fact that $\underline{Antp}$-like mutations are recessive lethals while $\underline{Ns}$ is viable as a homozygote. An additional head transforming mutation is resident at the locus and is called $\underline{Cephalothorax}$ ($\underline{Ctx}$) (Fig. 6), and it is also homozygous lethal. The basic effect caused by all three lesions is a transformation of elements derived from the eye-antennal disk into mesothoracic structures. Unlike $\underline{Scr}$ and $\underline{Ubx}$, a heterozygous deletion of the $\underline{Antp}$ locus does not yield the same dominant transformation. Indeed, there is no dominant phenotype associated with such deletions (Lewis et al., 1980b), and the dominant phenotype of $\underline{Ns}$ and $\underline{Antp}$ can be reverted by deleting the locus (Denell, 1973; Duncan and Kaufman 1975; Struhl, 1981; Hazelrigg, 1982). Therefore, the observed transformation results from the improper expression of the locus rather than its inactivity. The same conclusion applies to the $\underline{Extra}$ $\underline{Sex}$ $\underline{Combs}$ ($\underline{Scx}$) and $\underline{Humeral}$ ($\underline{Hu}$) mutations. The $\underline{Scx}$ lesion has the reciprocal effect of the $\underline{Scr}$ mutations described above. Sex combs are found on all six legs of adult males (Fig. 7). Other aspects of the chetotaxy of these legs indicates that the entire mesothoracic and anterior half of the metathoracic legs are transformed into a prothoracic identity. Like $\underline{Antp}$, $\underline{Scx}$ is associated with a recessive lethality and this lethality is shared by both lesions; i.e., $\underline{Antp}/\underline{Scx}$ heterozygotes are lethal and both fail to complement deletions of the $\underline{Antp}$ locus.

The $\underline{Hu}$ mutation results in a proliferation of bristles and hairs on elements of the dorsal prothorax (Fig. 8). This mutant phenotype is more extreme when $\underline{Hu}$ is made heterozygous for deletions of the $\underline{Antp}$ locus. Similar to $\underline{Antp}$ and $\underline{Ns}$, $\underline{Scx}$ and $\underline{Hu}$ can be reverted by deletion and, therefore, the dominant phenotypes produced result from an abnormal action of the $\underline{Antp}$ locus of which they are alleles.

A final class of mutations exists at the $\underline{Antp}$ locus. These are recessive lethals with no apparent dominant phenotypes. A majority of these fail to complement the recessive lethalities of $\underline{Antp}$ and $\underline{Scx}$ and show the same lethal phase and phenotype. The lethal phase is similar to that of the $\underline{Scr}$ mutations in the late embryo just prior to

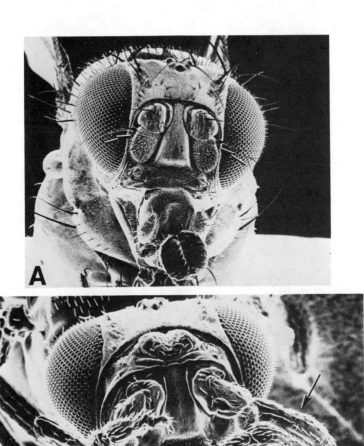

Figure 5: Scanning electron micrographs of a normal (A) and Antp^{73b}
(B) individual showing a frontal view of the head. Arrows in A
indicate the normal antennae; single arrow in B indicates the
leg-like structure which replaces the antennae.

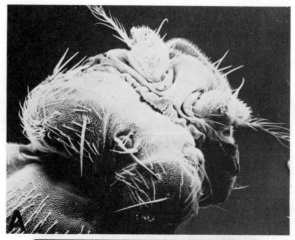

Figure 6: Scanning electron micrographs of the head of a
Cephalothorax (Ctx) mutant individual. A) Dorsal view. B) Lateral
view. Note the absence of an eye and the presence of dorsal
mesothoracic (T_2) tissue.

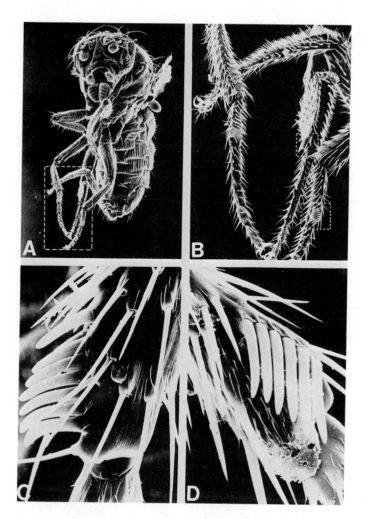

<u>Figure 7</u>: Scanning electron micrographs of an <u>Extra</u> <u>Sex</u> <u>Combs</u> (Scx)
individual. A) Ventral view of entire male animal. The dashed
line surrounds the distal portions of the meso- and metathoracic
legs. B) Enlarged view of the boxed area in A. The two small
dashed line boxes surround the sex combs on the meso- (bottom) and
metathoracic (top) legs. C) Close-up view of a mesothoracic sex
comb. D) Close-up view of a metathoracic sex comb. Other aspects
of the chetotaxy of these legs indicates that both the anterior and
posterior compartments of the mesothoracic leg are transformed to
prothorax while only the anterior compartment of the metathoracic
leg is transformed. The legs of females are also transformed.

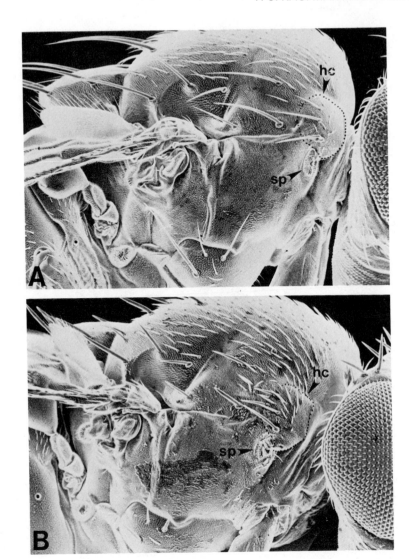

Figure 8: Scanning electron micrographs of a normal (A) and Humeral (Hu) (B) mutant individual showing the lateral (plueral) aspect of the thorax. The dotted line surrounds the humeral callus (hc) and the arrow below the callus shows the position of the mesothoracic spiracle. Note the presence of extra bristles and hairs on the callus and ventral to the spiracle in the Hu animal relative to normal. The structures affected (humeral callus and propluerae) are derived from the humeral or dorsal prothoracic imaginal disc.

hatching. The phenotype of these animals is shown in Fig. 9. In all cases the mesothoracic segment is transformed into the likeness of a prothorax ($T_2 \rightarrow T_1$). Occasionally, some animals exhibit a similar transformation of metathorax to prothorax ($T_3 \rightarrow T_1$). In addition to these transformations, small sclerotized plates are formed in the posterior portions of the prothoracic segment which is apparently normal anteriorly. These plates are reminiscent of the material that normally comprises the mouthparts and may be indicative of a transformation of the posterior T_1 into a gnathal segment. The phenotype of these lethal animals (T_3 and $T_2 \rightarrow T_1$) is similar to that seen in the Scx/+ adults, but there is no obvious effect or defect in the embryonic head.

In order to further investigate the relationship of the dominant gain-of-function lesions at the Antp locus and the recessive lethality, we induced clones of antp$^-$ cells at various times in development. As can be seen in Fig. 10, such clones develop normally in the antenna. Therefore, Antp$^+$ function is unecessary for normal antennal development. The same is not true of the legs. Clones induced after 24 hr of development during the larval stages exhibited a characteristic fusion of the tibia and femur. This was found for all three legs (T_1, T_2, and T_3). If the clones were induced prior to 24 hr during the embryonic period, we observed various transformation of proximal leg into proximal antenna in the mesothoracic but not pro- or metathoracic legs. These latter appendages still only showed the tibial/femoral fusion phenotype (Fig. 10). A similar result has been obtained using revertants of Ns (Struhl, 1981).

Based on these results it would appear that Antp is similar to Ubx in that it performs two similar but distinct functions in ontogeny. Early in development, Antp$^+$ is necessary to specify proper thoracic identity. In the absence of this function a T_1-like or gnathal identity replaces the normal T_1-posterior, T_2, and T_3 patterns. Later in development, during the stages in which Ubx$^+$ and Scr$^+$ are functioning to maintain the specified fates of the segments in which they are active, Antp$^+$ apparently functions to repress antennal development in the ventral portions of the three thoracic seqments.

This bifunctional model of Antp$^+$ activity is supported by a newly recovered lesion of the locus. This mutation (antp^{a58}) is a recessive lethal which does not cause death at the end of embryogenesis but allows the animals to survive until the late pupal stage. The prothoracic legs of two such animals are shown in Fig. 11. The morphology of these legs is strikingly similar to those obtained in the clonal analyses and possibly indicates a partial transformation of $T_1 \rightarrow An$. In any event, what is clear is that the early specification event (T_2 and T_3 versus T_1) takes place in these animals and the embryonic and larval stages are completely normal. Only the later maintenance function is disrupted.

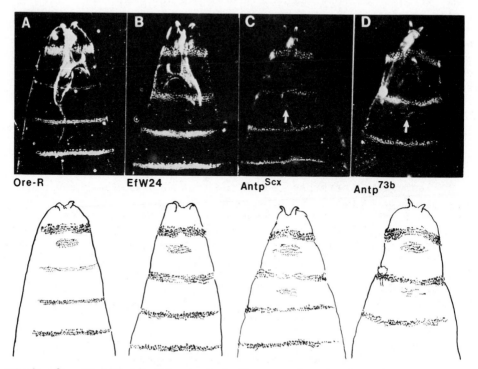

Figure 9: Darkfield illuminated micrographs showing the anterior
ventral cuticle of 24-hr-old embryos of a normal (A), an
Antennapedia recessive lethal (B) homozygote, an Antp^Scx/Df(3R)Scr
(C) individual, and an Antp^73b/Df(3R)Scr individual. Below are
tracings of the micrographs showing the position and morphology of
the ventral denticle belts. Note the thickened appearance of the
third row of denticles in the three mutant individuals relative to
normal. Also note the presence of a midventral patch of denticles
in the second thoracic (T_2) segment in C and D (indicated by
arrows). Some mutant individuals show a similar transformation of
the third thorax (T_3) to a prothoracic morphology. The posterior
portion of T_1 in all three mutant animals also bears a pair of
small sclerotized plates which are reminiscent of portions of the
mouth parts.

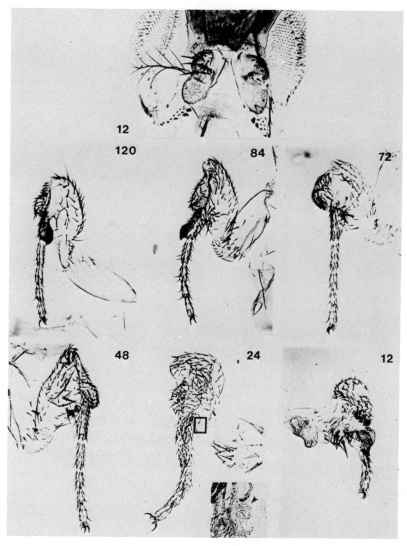

<u>Figure</u> 10: Photomicrographs of the head (top) and legs (bottom six
panels) of animals in which x-ray induced somatic clones have been
created in those structures. All of the clones are homozygous for
antp^{w10}, a recessive lethal allele at the <u>Antp</u> locus which normally
dies in the late embryo and expresses the morphology shown in Fig.
8. The numbers in each panel refer to the age of the animal in
which the clone was induced. The clone encompassing the entire
antenna (top) shows entirely normal morphology. Clones induced in
the legs show a characteristic fusion of the femur and tibia (see
Fig. 11A for normal morphology). This fusion takes place if the
clone is induced in the anterior or posterior compartments. Only
fusions are seen in pro- and metathoracic legs or in clones induced
after 24 hr in the mesothoracic legs. Clones induced in

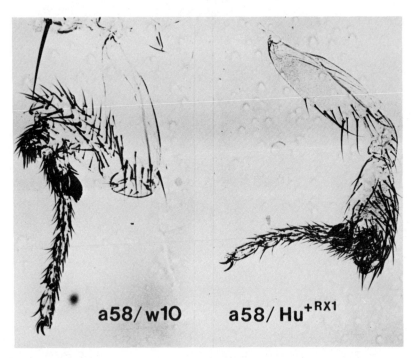

a58/w10   a58/Hu^{+RX1}

<u>Figure 11</u>: Photomicrographs of prothoracic legs of individuals
carrying the $\underline{antp}^{a58}$ recessive lethal mutation. There is a fusion
of femur and tibia similar to that seen in the clones of $\underline{antp}^-$
cells. Only the prothoracic legs are affected; the others are
normal as are all other adult structures. The phenotype is
expressed in heterozygous combination with either point mutations
(left) or deletions (right).

<u>Figure 10</u> (continued):

mesothoracic legs at or prior to 24 hr also show partial
transformations of leg to antenna. The boxed area in the 24-hr
panel indicates a region of the fused tibia/femur containing a
sacculus (a structure normally found only in the antenna). The
inset in this same panel is an enlargment of the boxed region. The
arrow indicates the sacculus. The 12-hr panel shows the maximum
leg → antenna transformation. There is often good transformation
of proximal structures, but we have not observed the distal tarsal
portions of the leg to transform to arista.

In complementation tests utilizing this newly recovered allele, it was discovered that another class of recessive mutations exist at the locus. The majority of recessive lethals at Antp fail to complement antp^{a58} and produce pharate adults with the previously described leg phenotype. A few lethals, however, were found which completely complement antp^{a58} and produce morphologically normal adult flies. To date, this group includes four apparent point mutations, a translocation, and a partial deletion of the locus Df(3R)A60 (Fig. 1). Developmental studies on these recessive lethal point mutations revealed that they, too, in some cases will survive until the pharate adult stage. However, they do not exhibit the partial $T_1 \rightarrow$ An transformation, but instead show a disruption in the development of the dorsal prothorax (Fig. 12). We are not yet certain whether their aberrant morphology is representative of a homoeotic transformation since there are no novel structures produced on the humeral excresences. However, if this dorsal disruption is analogous to those seen in the ventral structures, we might expect the transformation to be toward a head-like identity. Since a majority of the adult head is formed from ventral elements, such a transformation may be precluded and therefore no definitive structure is formed. This argument is, of course, less than compelling and further analyses will be necessary before any real conclusions can be formed. However, what is apparent is that we have found an additional functional complexity at the Antp locus and a clear case of intragenic complementation. This fact is further supported by a clonal analysis performed by one of us (M.A.) using the antp^{r4} mutation which belongs to that set of lesions which complement antp^{a58} and produce humeral outgrowths. The antp^{r4} clones are completely normal in all three legs and in any adult structure in which they are found. This result, coupled with the morphology of the pharate adults (Fig. 12) and the complementation results, makes it apparent that it is possible to independently disrupt the early and late functions of the Antp locus.

This latter point is supported by our clonal analyses of two of the dominant gain-of-function lesions at the locus. Using an EMS-induced Scx mutation, Wakimoto (1981) was able to induce, recover, and characterize somatic clones in all adult structures, except those derived from the humeral disk where they apparently do not survive. Several of these clones are shown in Fig. 13. In no case was the leg morphology observed with antp^{w10} (Fig. 10) clones seen in clones homozygous for Scx^W. The only transformations were similar to those found in Scx^W/+ heterozygotes; i.e., T_2 and $T_3 \rightarrow T_1$. It was also found that clones in the prothoracic leg and labial structures showed partial transformations of $T_1 \rightarrow T_2$ and lab $\rightarrow$ Mx (Fig. 14), respectively. These two transformations are similar to those seen in animals carrying lesions at the Scr locus, and their presence in these clones is consistant with genetic data which indicates that the Scx^W lesion results from a simultaneous affect on

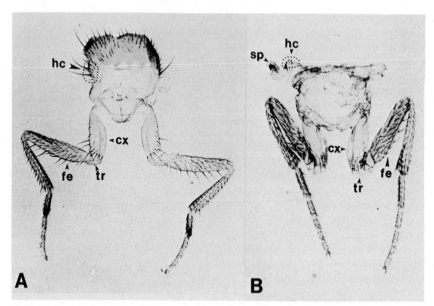

Figure 12: Photomicrographs of the frontal view of the dorsal and
ventral prothorax of a normal (A) and an antp^{r10}/antp^{w24} (B)
individual. The head has been removed in both cases. Note the
normal morphology of the legs and the outgrowths in the dorsal
prothorax. sp = spiricle, hc = humeral callus, cx = coxa, tr =
trochanter, fe = femur

the Scr and Antp loci. Regardless of their genetic complexity, it is
clear that the Scx^W mutation does not affect the late maintenance
function of Antp$^+$ since the leg clones are normal legs not antennae.
This is the case despite the fact that Scx^W/Df(3)Scr animals die as
late embryos exhibiting a characteristic antp$^-$ phenotype (Fig. 9;
Wakimoto and Kaufman, 1981).

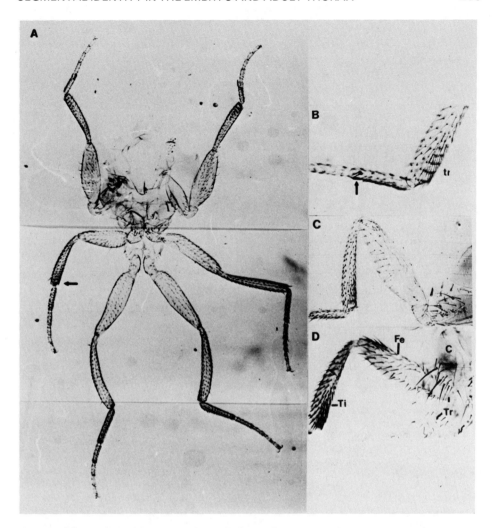

Figure 13: Photomicrographs of legs bearing clones of cells
   homozygous for the Extra Sex Combs of Wakimoto (Scx^W) mutation. A)
   Entire set of legs with a clone filling the entire anterior
   compartment of the mesothoracic leg (arrow). Note the presence of
   transverse rows of bristles on the tibia; these are normally only
   found on the prothoracic leg. B) Clone in the anterior compartment
   of the prothoracic leg. There are transverse rows (tr) on the
   tibia but there is also a severe reduction in the number of sex
   comb teeth (arrow). C) Clone in the anterior compartment of the
   mesothoracic leg. D) Clone in the anterior compartment of the
   metathoracic leg. Note the normal morphology of the femur (Fe) and
   tibia (Ti). C = coxa, Tr = trochanter

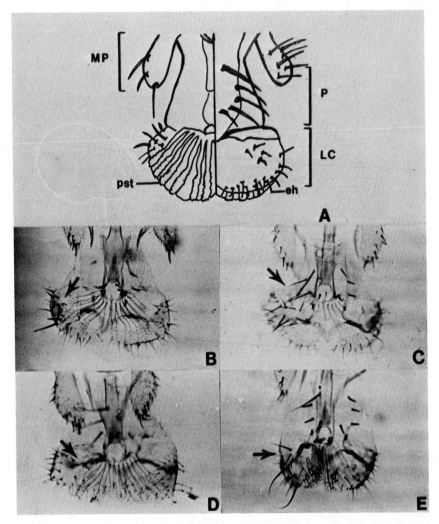

Figure 14: Drawing of the anterior (left) and posterior (right)
aspects of the mouth parts (A) and photomicrographs (B, C, D and E)
of the mouth parts of animals bearing clones of cells homozygous
for the Scx^W mutation. Note the disruptions of the psuedotrachea
(pst) (arrows in B and D) and the elaboration of extra bristles on
the prementum (P) (arrow in C) and the labial palps (LC) in the
region of the edge hairs (eh) (arrow in E). These transformations
are reminiscent of those seen in Scr lesion-bearing individuals
that survive to adulthood (see Fig. 3D). MP = maxillary palp

$Antp^{73b}$, like Scx^W, shows the early $antp^-$ lethal phase and phenotype (Fig. 9; Wakimoto and Kaufman 1981). Homozygous $Antp^{73b}$ somatic clones, like those for Scx^W, survive in a majority of the adult cuticle and only express a more extreme version of their heterozygous phenotype. There is no indication of a leg → antenna transformation. Therefore, $Antp^{73b}$, like Scx^W (and by inference similar Antp gain-of-function lesions), disrupts the late maintenence function of $Antp^+$. However, both alter the early embryonic function which is necessary for thoracic specification.

Interactions between Scr and Antp

The Scr and Antp loci are apparently adjacent to each other in the ANT-C (Fig. 1). This juxtaposition may be of some significance since, when mutated, they produce reciprocal thoracic transformations during embryogenesis. Scr mutations result in the prothorax to mesothorax (T_1 → T_2) transformation while Antp nulls transform T_2 to T_1. We have constructed, by recombination, a double mutant chromosome (i.e., Scr^{w17} $antp^{w10}$) in order to determine if either transformation would predominate or if some novel transformation might result. As was expected, animals homozygous for this chromosome die late in embryogenesis. The phenotype of these animals is not dissimilar to animals carrying only the $antp^{w10}$ mutation (see Fig. 15). Therefore, we can conclude that Antp is epistatic to Scr and that a deficiency of Scr is only revealed in an animal which has a functional Antp locus. Put another way, Scr function in the thorax is in some manner dependent on normal Antp function. The same statement cannot be made for Scr function in the labial segment. In the case of either the single or double mutant chromosome, the labium is transformed to maxilla.

A similar result for the mouthparts was obtained by Struhl (1982) from a clonal analysis of an Scr^- $antp^-$ double mutation. However, he also showed that $Antp^+$ function in the prothorax is modulated by Scr. Recall that Scr^+ $antp^-$ clones partially transform the mesothoracic leg to antenna but only cause the femoral-tibial fusions in the prothoracic legs. Clones which are Scr^- $antp^-$ cause the leg → antenna transformations in both T_1 and T_2 (Fig. 15). Therefore, $antp^-$ cells are apparently prohibited from completely transforming ventral T_1 to antenna by Scr^+, and removal of this restriction allows this transformation to occur. It follows that in early development, $Antp^+$ function in T_1 can be viewed as a prerequisite to Scr^+ function while later in development, Scr^+ acts in turn to modify $Antp^+$ function.

Double Mutants of Antp and Ubx

Since $antp^-$ individuals show a partial transformation of T_3 to T_1 and Ubx^- individuals transform T_3 and A_1 to T_2 (Fig. 15), we decided to look for interaction between these two loci as well. To

Scr	Antp	Ubx		An	Mx	Lb	T₁	T₂	T₃	A₁
+	+	+		An	Mx	Lb	T₁	T₂	T₃	A₁
−	+	+		An	Mx	Mx	T₂	T₂	T₃	A₁
+	−	+ E		An	Mx	Lb	T₁	gn	T₁	T₃	A₁
+	−	+ L		An	Mx	Lb	An	An	An	A₁
+	+	− E		An	Mx	Lb	T₁	T₂		
+	+	− L		An	Mx	Lb	T₁	T₂	T₂	T₂
−	−	+ E		An	Mx	Mx	T₁	gn		A₁
−	−	+ L		An	Mx	Mx	An	An	An	A₁
−	+	−		An	Mx	Mx	T₂	T₂	T₂	
+	−	−		An	Mx	Lb	T₁	gn		
−	−	− E		An	Mx	Mx	T₁	gn	gn	T₁
−	−	− L		An	Mx	Mx	An	An	An	?

			Scr⁺ Antp⁺	Antp⁺	Ubx⁺ Antp⁺	Ubx⁺
Scr⁺						
An	Mx	Lb	T₁	T₂	T₃	A₁

Figure 15: Summary of segmental transformations seen in animals
deficient for Scr, Antp, and Ubx functions. The pluses and minuses
under the locus designations at the left indicate the genotypic
state for the figured segmental transforms to the right. The E and
L designations indicate differences in the early vs. late functions
of the genes in question. Inside each boxed segment is listed the
identity of that segment. The crosshatching indicates segments or
parts of segments which are transformed. Left slanting lines show
transformations to a more anterior identity; right slanting a more
posterior one. The bottom-most portion of the figure indicates the
segmental domains in which the three loci are active. An =
antenna, "An" = antennal transformation incomplete, Mx = maxillary,
Lb = labial, T_1 = prothorax, "T_1" = prothoracic transformation
variable, T_2 = mesothorax, T_3 = metathorax, A_1 = first abdominal,
gn = gnathal-mouthpart-like structures are found but their identity
is uncertain

this end, $antp^{w10}$ Ubx^1 double mutants were constructed and their
lethal phase and morphology determined. These animals die in the
late embryo and show a transformation of T_1 and T_2 to the mixed
T_1/gnathal phenotype observed in antp⁻ animals. Additionally, the T_3
and A_1 segments are transformed into a T_1-like identity. Thus,
concomitant deletion of Antp and Ubx gives a thoracic identity
(i.e., T_1-like) over the entire group of segments having a T_2-like
thoracic identity in Ubx⁻ individuals. Therefore, in terms of
segmental identity, Antp is epistatic to Ubx, but Antp can be

extended beyond its normal range of effects into the A_1 segment by loss of the Ubx^+ function.

Triple Mutants Scr⁻ Antp⁻ Ubx⁻

Finally, we have constructed a chromosome carrying lesions in all three of the thorax specifying loci. Animals homozygous for this chromosome die late in embryogenesis and have the phenotype shown in Fig. 15. The labial segment is transformed to maxillary as in Scr⁻ individuals. All three thoracic segments are transformed into the mixed T_1/gnathal phenotype seen in antp⁻ prothoraces, and the A_1 segment is T_1-like. The major differences between these individuals and the Antp Ubx double mutants is in the Lb → Mx transformation and the formation of mouthpart-like structures on the posterior of the normal T_3 segment.

Taken together, all of the above results show the functional domains and interactions of the three major loci involved in the specification and maintenance of thoracic identity of Drosophila. In the two segments flanking the thorax proper, the labial and first abdominal segments are specified by the Scr and Ubx loci. It is interesting to note that both of these genes also function in thoracic specification. It would appear that Antp is the central figure in the specification of thoracic identity. Mutations at this locus affect all three segments while Scr and Ubx only affect T_1 and T_3, respectively. Moreover, multiple mutations involving Antp and the other two loci always produce abnormal morphologies or transformations characteristic of the Antp locus lesions. However, it is also apparent that Scr, Antp, and Ubx are acting in concert and in a highly orchestrated manner both spatially and temporally to achieve the normal pattern of thoracic development. The elucidation of the precise manner of their action, interaction, and orchestration must await further experimentation.

The Nature of Lesions at the Antp Locus

Despite our inability to critically address the precise mode of action of the various homoeotics discussed above, our current genetic and developmental data have allowed us some insight into the nature of the genetic lesions at the Antp locus. As we have, shown there are at least two complementing functions at the Antp locus. The first of these is necessary in early development and serves to select among or specify the various thoracic types seen in the larva. The second of these is required later in development to maintain the specified thoracic state. Most Antp recessive lethals disrupt both of these functions, as is shown by their early lethal phenotype and the morphology of somatic clones of antp⁻ cells. This result is shown diagramatically in Fig. 16. Some recessive lethals, on the other hand, affect one or the other of these two functions: $antp^{a58}$ is only defective in the late maintenance function while Df(3R)A60

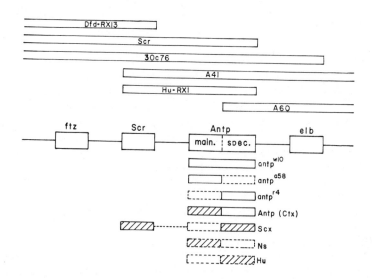

Figure 16: Diagramatic representation of the various lesions at the
Antp locus also showing the flanking ftz, Scr, and elb loci (see
Fig. 1 for abbreviations). The open boxes above the loci indicate
the area of the Antp locus. The locus is divided into two
functional components: a more proximal maintenance (main.)
function and a more distal specification (spec.) function. The
bars below the locus indicate which of these two functions are
disrupted by the lesions listed to the right of each bar. A solid
line open bar indicates disruption (inactivation). A dashed open
bar indicates no disruption. Crosshatching indicates an abnormal,
spatial, or temporal expression (regulation) of the function in
question.

and, by inference, antp^{r4} disrupt only the specification function
(Fig. 16). The dominant Antp lesions are all defective for the early
specification function and thus die as late embryos showing a T$_2$ and
T$_3$ to T$_1$ transformation. However, they are not defective for the
late maintenance function in the normal thoracic legs (Antp^{73b} clones
are normal in the legs). But they do "misregulate" this function.
Late in development, Antp$^+$ can be viewed as a suppressor of antennal
development in the imaginal disks. If this pathway is suppressed,
leg development occurs simply by default. The Antp dominant lesions
apparently express the Antp$^+$ repressive function in the antennae
where it is not normally expressed, thereby allowing legs to develop.

The Scr lesions are like the Antp mutations: blocked in the
early function but not in the late function. They, too, show a
misregulation, but of a different sort. Since the antp$^-$ embryos show

a T_2 and T_3 to T_1 transformation as do the Scx/+ adults we view this misregulation as a temporal one. The normally early specification function is now expressed later in development and appears (albeit in mutant form) in the adult. Since the mutation we have analyzed also carries a lesion at the Scr locus, we cannot rule out that some of the observed effects of this lesion are due to Scr and/or an Scr-Antp interaction. Indeed, the latter explanation is quite probable and may account for the novel phenotype of this mutation relative to most dominant gain-of-function alleles.

Finally, it should be noted that both of the proposed functions of Antp can be misregulated without causing a lethal disruption of the locus. The Ns mutation represents a non-lethal, spatially abnormal expression of the $Antp^+$ maintenance function, and the Hu phenotype may result from an abnormal expression of the function necessary for normal dorsal and lateral prothoracic development.

Prospectus

Much of the genetic and developmental work described here has resulted in interpretations which are essentially phenomenological. While the experiments have allowed us to define the apparent times and places of action of the various homoeotic loci and has, to some extent, constrained the number of hypotheses permissible to explain the primary function(s) of these loci, no direct proof exists as to the proximate cause of the observed transformations or of the nature and function of these gene products.

To expand the investigation of these loci, both the BX-C and the ANT-C are currently being analyzed at the molecular level utilizing recombinant DNA technology. While we do not believe that these cloned genomic fragments and their further characterization will, in themselves, produce definitive answers to questions about how these genes and their products perform in normal ontogeny, we do believe that the molecular analysis, in combination with further genetic and developmental experimentation, will produce greater insight into the function of these intriguing genetic entities.

ACKNOWLEDGEMENTS

The authors would like to acknowledge the support and contributions of their colleagues in the fly lab: for their contributions to the genetic and developmental work, Drs. T. Hazelrigg, R. Lewis, and B. Wakimoto; for the SEMs used in this paper, Dr. F.R. Turner; for valuable discussions, Dr. M. Scott, Ms. A. Weiner, Ms. L. Cain, and Mr. M. Morton; for technical expertise Ms. P. Fornili and Ms. D. Otteson. Supported by PHS-GM24299 and PHS-GM29709.

REFERENCES

Becker, H.J. (1975). Mitotic Recombination. In "The Genetics and Biology of Drosophila," Vol. 1c, Academic Press.

Bennett, D. (1975). The T-locus of the mouse. Cell 6:441-454.

Capdevila, M.P. and Garcia-Bellido, A. (1978). Phenocopies of bithorax mutants. Wilhelm Roux' Archiv. 185:105-126.

Chalfie, M., Horvitz, R., and Sulston, J. (1981). Mutations that lead to reiterations in the cell lineages of C. elegans. Cell 24:59-69.

Denell, R.E. (1973). Homeosis in Drosophila. I. Complementation studies with revertants of Nasobemia. Genetics 75:279-297.

Denell, R.E., Humels, K.R., Wakimoto, B.T., and Kaufman, T.C. (1981). Developmental studies of lethality associated with the Antennapedia gene complex in Drosophila melanogaster. Devel. Biol. 81:43-50.

Duncan, I.W. and Kaufman, T.C. (1975). Cytogenetic analysis of chromosome 3 in Drosophila melanogaster: Mapping of the proximal portion of the right arm. Genetics 80:733-752.

Garcia-Bellido, A. (1977). Homoeotic and atavic mutation in insects. Amer. Zool. 17:613-630.

Garcia-Bellido, A. and Ripoll, P. (1978). Cell lineage and differentiation in Drosophila. In "Results and Problems in Cell Differentiation," Vol. 9:119-156. Springer-Verlag, Berlin.

Gehring, W. (1966). Bildung eines vollstandigen Mittelbeiner mit Sternopleura in der Antennenuregion bei der Mutante Nasobernia (Ns) von Drosophila melanogaster. Arch. Julius Klaus Stift. Vererbungrforsch. Sozialanthropol. Rossenburg. 41:44-54.

Hazelrigg, Tulle I. (1982). Cytogenetic properties and developmental function of selected portions of the Drosophila genome. Ph.D. dissertation, Department of Biology, Indiana University.

Kauffman, S.A. (1975). Control circuits for determination and transdetermination: Interpreting positional information in a binary epigenetic code, pp. 201-214. In "Cell Patterning," Ciba Foundation Symposium 29, Elservier.

Kauffman, S.A. (1977). Characteristic waves compartments and binary decisions in Drosophila development. Amer. Zool. 17:631-648.

Kauffman, S.A. and Ling, E. (1980). Timing and heritability of the Nasobemia transformation in Drosophila. Wilhelm Roux' Archiv. 189:147-153.

Kauffman, S., Shymko, R., and Trabert, K. (1978). Control of sequential compartment formation in Drosophila. Science 199:259-270.

Kaufman, T.C. (1978). Cytogenetic analysis of chromosome 3 in Drosophila melanogaster: Isolation and characterization of four new alleles of the proboscipedia (pb) locus. Genetics 90:579-596.

Kaufman, T.C., Lewis, R.A., and Wakimoto, B.T. (1980). Cytogenetic analysis of chromosome 3 in Drosophila melanogaster: The homoeotic gene complex in polytene chromosome interval 84A-B. Genetics 94:115-133.

Kaufman, T.C. and Wakimoto, B.T. (1982). Genes that control high level developmental switches. In "Evolution and Development," J.T. Bonner (ed.), pp. 189-205. Dahlem Konferenzen, 1982.

Kerridge S. and Morata, G. (1982). Developmental effects of some newly induced Ultrabithorax alleles of Drosophila. J. Embryol. Exp. Morph. 68:211-234.

Lewis, E.B. (1963). Genes and developmental pathways. Am. Zoologist 3:33-56.

Lewis, E.B. (1964). Genetic control and regulation of developmental pathways. In "Role of Chromosomes in Development," pp. 231-252. Academic Press.

Lewis, E.B. (1978). A gene complex controlling segmentation in Drosophila. Nature 276:565-570.

Lewis, E.B. (1981a). Develpmental genetics of the Bithorax complex in Drosophila. In "Developmental Biology Using Purified Genes," D.D. Brown and C.F. Fox (eds.), ICN-UCLA Symposium on Molecular and Cellular Biology, Vol. XXIII. Academic Press, New York.

Lewis, E.B. (1981b). Control of body segment differentiation in Drosophila by the Bithorax gene complex. In "Embryonic Development: Genes and Cells," M.M. Burger (ed.), Proceedings of the IX Congress of the International Society of Developmental Biologists. Alan Liss, Inc., New York, in press.

Lewis, R.A., Kaufman, T.C., Denell, R.E., and Tallerico, P. (1980a). Genetic analysis of the Antennapedia gene complex (ANT-C) and adjacent chromosomal regions of Drosophila melangaster. I. Polytene chromosome segments 84B-D. Genetics 95:367-381.

Lewis, R., Wakimoto, B., Denell, R., and Kaufman, T. (1980b).
    Genetic analysis of the Antennapedia gene complex (ANT-C) and
    adjacent chromosomal regions of Drosophila melanogaster. II.
    Polytene chromosome segments 84A-84B1,2. Genetics 95:383-397.

Minana, F.J. and Garcia-Bellido, A. (1982). Preblastoderm mosaics
    of mutants of the Bithorax complex. Wilhelm Roux' Archiv.
    191:331-334.

Morata G. and Kerridge, S. (1981). Sequential functions of the
    Bithorax complex of Drosophila. Nature 290:778-781

Ouweneel, W. (1976). Developmental genetics of homoeosis. Adv.
    Genet. 18:179-248.

Schubiger, G. and Newman, Jr., S.M. (1982). Determination in
    Drosophila embryos. Amer. Zool. 22:47-55.

Schubiger, G. and Wood, W. (1977). Determination during early
    embryogenesis in Drosophila melanogaster. Amer. Zool.
    17:565-576.

Struhl, G. (1981). A homoeotic mutation transforming leg to antenna
    in Drosophila. Nature 292:635-638.

Struhl, G. (1982). Genes controlling segmental specification in the
    Drosophila thorax. Proc. Natl. Acad. Sci. USA 79:7380-7384.

Wakimoto, B.T. (1981). Genetic and developmental studies of the
    loci in the polytene interval 84A-84B1,2 of Drosophila
    melanogaster. Ph.D. dissertation, Department of Biology,
    Indiana University.

Wakimoto, B.T. and Kaufman, T.C. (1981). Analysis of larval
    segmentation in lethal genotypes associated with the
    Antennapedia gene complex in Drosophila melanogaster. Devel.
    Biol. 81:51-64.

ISOLATION AND CHARACTERIZATION OF GENES DIFFERENTIALLY

EXPRESSED IN EARLY <u>DROSOPHILA</u> EMBRYOGENESIS

Judith A. Lengyel, Steven R. Thomas,
Paul David Boyer, Fidel Salas, Teresa R. Strecker,
Inyong Lee, Melissa L. Graham, Margaret Roark,
and Eileen M. Underwood*

ABSTRACT

Important events of <u>Drosophila</u> embryogenesis are compressed into a remarkably short period: in the first 3.5 hours after fertilization, rapid nuclear multiplication, cellularization, activation of transcription, and the determination of much of the basic pattern of the larva and adult occur. In this chapter we review the evidence that only a relatively small number of genes are involved in controlling these processes, and that of such genes, those transcribed as part of the maternal program may provide qualitatively different information than those transcribed as part of the zygotic program. Reasoning that genes which play a major role in events specific to early embryogenesis should be expressed differentially during this time period, we have used recombinant DNA techniques to isolate members of two such classes of genes. One class encodes mRNAs which are part of the maternal population, but which disappear rapidly during embryogenesis; the other encodes mRNAs whose expression is primarily zygotic, reaching a peak at the blastoderm or late gastrula stage. We expect that genetic and bio-chemical analyses of these cloned sequences will provide insight into events specific to early embryogenesis.

*From the Molecular Biology Institute and Department of Biology, University of California, Los Angeles, Los Angeles, CA 90024.

INTRODUCTION

The first few hours of embryogenesis in Drosophila are crucial
in setting the stage for later events of morphogenesis and
cytodifferentiation. In this brief time, nuclei replicate at a
maximal rate, cells are formed, and RNA transcription is dramatically
activated. Of greatest importance during this period is the
determination of cells and tissues which establishes the basic body
plan of larva and adult.

Only a small number of genes appear to be involved in
controlling events of early embryogenesis. As reviewed below
(Sections II and III), we expect that the expression of many such
controlling genes is restricted to this early time period, and that
genes specific to the maternal program (transcribed only during
oogenesis) provide a different class of information than those
specific to the zygotic program (newly activated when transcription
begins in the embryo). While both expectations appear to have been
met for a very few genes known to affect events specific to early
embryogenesis, it is generally difficult, on the basis of purely
genetic criteria, to identify genes whose activity is restricted to a
particular developmental period.

Our approach has been to use molecular techniques to identify
genes which make a specific maternal or zygotic mRNA contribution to
the first few hours of Drosophila embryogenesis. We expect the
analysis of such genes to provide an important complement to analysis
of genes affecting early embryogenesis that have been identified by
genetic techniques. The characterization of both types of genes will
give insight into a number of questions. How many genes control
early embryogenesis? What is the function of such genes? Do genes
specific to the maternal or zygotic program have qualitatively
different functions in early embryogenesis?

Molecular and Cellular Events During Early Drosophila Embryogenesis

The events of early embryogenesis in Drosophila have been well
characterized at the morphological, genetic, and biochemical level.
A recent review of the morphology of Drosophila embryogenesis has
been presented by Fullilove and Jacobson (1978).

A. Nuclear replication and cellularization. Immediately
following fusion of the male and female pronuclei (15 min after
oviposition at 25°C) there are nine synchronous nuclear divisions
(Sonnenblick, 1950); this is referred to as the syncytial cleavage
stage (0.3-1.6 hr) because there is no subdivision of the embryo into

cells at this time. During this period the nuclei divide every 9-10 min. They then migrate to the surface of the embryo where they undergo four more rapid synchronous cleavages, with the cycle time gradually increasing to 18 min (Warn and Magrath, 1982; Zalokar and Erk, 1976); this is referred to as the syncytial blastoderm stage (1.6-2.5 hr). The rate of nuclear multiplication during these syncytial stages is the most rapid for any eucaryote, with an S phase of only 3-4 min (Rabinowitz, 1941). Following the synchronous divisions at the syncytial blastoderm stage, cellularization occurs (2.5-3.5 hr) at which time there is a lull in nuclear multiplication (Madhavan and Schneiderman, 1977). Membranes from the egg surface move downward into the embryo and surround the embryonic nuclei. This is the cellular blastoderm stage and lasts until 3.5 hr, when gastrulation begins.

The fact that most known mutants affecting nuclear multiplication and cellularization are maternal effect mutants (Bakken, 1973; Rice and Garen, 1975; Swanson and Poodry, 1980; Thierry-Mieg, 1976; Zalokar et al., 1975), and that little transcription occurs prior to these events, suggests that these events are controlled largely by the maternal genome. There are, however, a few zygotic lethal mutants which also affect these processes (Wright, 1970).

B. Maternal versus zygotic mRNA. Most of the mRNA mass and the great majority of the mRNA sequence complexity needed for development to the gastrula stage are stored in the oocyte. There are approximately 6000 different species of mRNA of average length 2000 nucleotides stored in mature (stage 14b) oocytes (Hough-Evans et al., 1980); most of the sequence complexity of this mRNA population is present throughout embryogenesis (Arthur et al., 1979). Transcription from the embryonic genome becomes detectable at syncytial blastoderm and reaches its highest embryonic rate, on a per nucleus basis, at the cellular blastoderm. This has been demonstrated by autoradiography of individual embryos, electron microscopy of chromatin spreads, and biochemical analysis of embryo populations (Zalokar, 1976; McKnight and Miller, 1976; Anderson and Lengyel, 1979, 1981). Evidence for transcription of some nuclear genes during the syncytial cleavage stage, however, comes from the arrest of a number of zygotic lethal mutants at this stage (reviewed by Wright, 1970) as well as the isolation of cloned sequences homologous to mRNAs newly expressed at this stage (Sina and Pellegrini, 1982).

At the start of gastrulation, sufficient embryonic transcripts have accumulated to constitute 14% of the mass of mRNA on polysomes (Anderson and Lengyel, 1981). Thus both maternal mRNAs and new embryonic transcripts have the potential to control events during early embryogenesis.

i) <u>Maternal</u> <u>mRNA</u>. That some maternal genes make a unique contribution to embryogenesis is suggested by the existence of maternal effect mutations that survive as homozygous adults but produce defective progeny (Wright, 1970; King and Mohler, 1975; Bakken, 1973; Gans et al., 1975; Rice and Garen, 1975; Zalokar et al., 1976; Nusslein-Volhard, 1979). The normal appearance of the homozygous adults suggests that such genes are transcribed only during oogenesis. While it is difficult to rule out the possibilty that some maternal effect genes are transcribed at a very low level at other stages, maternal effect genes are known, such as those which affect embryonic axis polarity (Section IIIb), for which it is reasonable to conclude that transcription is limited to oogenesis.

Evidence for the existence of maternal mRNAs whose expression is unique to early embryogenesis also comes from molecular hybridization analyses. Hybridization enrichment experiments suggest that 2.5% of the cDNA transcribed from poly(A)$^+$ RNA of the mature oocyte does not hybridize to RNA of the 3.5-hr embryo (Arthur et al., 1979).

An important example of maternal information providing unique cytoplasmic determinants is the posterior polar plasm of <u>Drosophila</u>. Transplantation of posterior polar plasm from late oocytes or early embryos can induce the ectopic formation of functional germ cell precursors (Illmensee and Mahowald, 1974, 1976; Illmensee et al., 1976) and can correct UV-induced sterility (Okada et al., 1974). Similar posterior cytoplasm from older embryos cannot correct UV-induced sterility (Okada, personal communication). Cytochemically, RNA has been detected in association with the polar granules (electron dense structures located exclusively in the posterior polar plasm) in the early embryo, but it is no longer detected at later stages. This RNA may be associated with polysomes (Mahowald, 1971). Thus there may be maternal RNA (possibly an mRNA) stored in the posterior of the egg which functions early in development in germ cell determination and then decays.

The evidence from certain maternal effect mutations, from hybridization analyses, and from studies on polar granules thus suggests that there are genes transcribed only during oogenesis that supply RNAs to the early embryo that are not replaced by transcription from the zygotic genome. We expect these genes to exert an important influence over events of early embryogenesis. A definitive analysis of the temporal expression of such genes awaits the availability of recombinant DNA probes.

ii) <u>Zygotic</u> <u>transcription</u>. Both genetic and molecular evidence suggests that there are a number of genes which are not expressed in the maternal mRNA, but are newly expressed at the blastoderm stage, when the bulk of embryonic transcription has been shown to begin. Several zygotic lethal genes which affect segmentation must be newly expressed at the blastoderm stage (Section IIIc). A small but

significant fraction (8%) of the cDNA transcribed from blastoderm
mRNA can hybridize only with blastoderm but not with oocyte mRNA
(Arthur et al., 1979). Further hybridization analyses suggest that
there are approximately 40 of these mRNA sequences which are present
in the blastoderm embryo, but not in the oocyte (Goldstein and
Arthur, 1979). A screen for sequences differentially expressed in
2-5 hr embryos yielded only five different unique sequences (Scherer
et al., 1981). Since mRNAs which are continuously present during
embryogenesis are probably involved in functions common to all cells,
genes newly expressed at blastoderm are likely to be involved in the
critical events occurring at blastoderm: cellularization,
determination, and the beginning of morphogenetic movements.

C. _Protein synthesis_. While DNA and RNA synthesis increases
dramatically during the first few hours following oviposition,
activation of protein synthesis appears to occur earlier, during the
process of oviposition (Goldstein and Synder, 1973). After
oviposition there does not appear to be an increase in protein
synthesis (Goldstein and Snyder, 1973; Zalokar, 1976; Santon and
Pellegrini, 1983) and the fraction of total mRNA in polysomes
increases only slightly from 50% to 70% (Lovett and Goldstein, 1977;
Anderson and Lengyel, 1983). Significant changes in translation
efficiency may occur for some mRNAs, however; there is a three-fold
increase in polysome association of histone mRNA during the first 4
hr of embryogenesis (Anderson and Lengyel, 1983).

Are the changes in expression of a small number of genes during
early embryogenesis reflected by changes in the pattern of protein
synthesis? Of 300-400 resolvable proteins, less than 20 have been
identified that have a pattern of synthesis consistent with
translation only from transiently present maternal mRNA or only from
newly synthesized zygotic mRNA (Sakoyama and Okubo, 1981; Gutzeit,
1980; Savoini et al., 1981). Changes in protein synthesis in early
Drosophila embryos, once the process of translation has been
activated, thus seem to be restricted to a small number of species.
This is consistent with the conclusion, from genetic analyses and
studies on mRNA populations, that the total number of genes whose
activity changes during this period is small. This small group of
genes probably plays a crucial role in early embryogenesis.

Establishment of the Embryonic Pattern

A. _Experimental evidence for determination at blastoderm_. A
detailed fate map of the surface of the blastoderm has been
constructed from histological observations on cells during
gastrulation and primary organogenesis (Poulsen, 1950). Fate maps of
both larval and adult cuticular and internal structures have been
superimposed on this map by means of gynandromorph fate mapping
(reviewed by Janning, 1978). That a fate map can be constructed on
the surface of the blastoderm does not prove that the cells are

determined at this stage, but it does provide a framework for analysis of the state of determination of different cells in the surface of the blastoderm.

The results of genetic analyses, and ablation and transplantation experiments can all be interpreted as indicating that cells in the surface of the blastoderm are determined to give rise to ectodermal structures of the larva and adult. The developmental fate of transplanted anterior and posterior halves of blastoderm embryos indicates that anterior-posterior determination has already occurred at the blastoderm stage (Chan and Gehring, 1971). The behavior of X-ray induced marked clones is consistent with the idea that determination of segments occurs at the cellular blastoderm stage (Wieschaus and Gehring, 1976). That phenocopies of bithorax, a gene thought to affect segmentation, can be maximally induced at the blastoderm stage is also consistent with the idea that determination of segments occurs at this stage (Capdevila et al., 1974). Ablation or UV irradiation of patches of cells at the blastoderm stage, followed by observation of defects at the embryonic segmentation and larval stages, has allowed maps of the determined cells to be constructed which are consistent with each other and also with the blastoderm fate map constructed by histological techniques (Lohs-Schardin et al., 1979a, 1979b; Underwood et al., 1980). Transplantation of small groups of blastoderm cells into ectopic locations and analysis of adult structures resulting from donor tissue also indicate that blastoderm cells are determined (Simcox and Sang, 1983). Thus cells on the surface of the blastoderm are not only fated, on the basis of their position, to give rise to particular larval and adult structures, they also appear to be determined to become specific anterior or posterior compartments of the segments which will give rise to these structures.

B. Maternal contributions to the embryonic pattern. A number of maternal effect lethal mutations have been identified which affect the pattern of the embryo. These include bicaudal and dicephalic which affect the anterior-posterior axis (Nusslein-Volhard, 1979; Lohs-Schardin, 1982), a group of 14 mutants including dorsal, easter, gastrulation defective, and Toll which affect the dorsal-ventral axis and gastrulation (Nusslein-Volhard, 1979; Anderson and Nusslein-Volhard, 1983, and personal communication; Konrad et al., 1982), and esc which affects determination of embryonic segment identity (Struhl and Brower, 1982). The characteristics of these mutants have been used to support the hypothesis that maternal effect genes controlling embryo pattern act primarily to establish anterior-posterior and dorsal-ventral axes (possibly by concentration gradients) in the unfertilized egg (Nusslein-Volhard, 1979).

Is it possible that morphogenetic information might be present in the embryo in the form of maternal mRNA? We have already discussed the germ cell-determining posterior polar plasm which contains polar granules associated with RNA. While there is no evidence for the existence of other RNA species involved in determination in Drosophila, the involvement of maternal RNA in determination of embryonic axis polarity has been suggested by experimental manipulations in another dipteran, Smittia (reviewed by Kalthoff, 1979). Mirror image double abdomen embryos can be induced by a number of experimental techniques including centrifugation, anterior puncture, application of RNase to the anterior pole region and UV irradiation of the anterior pole, which is photo-reversible. The latter two cases point to the "anterior determinants" acting at least in part through RNA. Since experimental manipulations were carried out at a stage prior to the beginning of significant embryonic transcription (Jackle and Kalthoff, 1979), this RNA is probably maternal in origin.

C. Zygotic contributions to embryonic pattern. There is strong evidence that early transcription of a number of zygotic genes is involved in establishing the pattern of the Drosophila embryo. A group of 15 embryonic lethal complementation groups affecting segment determination (which is believed to occur at blastoderm) has been isolated. Three of these mutants, runt, knirps, and Kruppel, show an abnormal phenotype shortly after the onset of gastrulation (Nusslein-Volhard and Wieschaus, 1980); the genes defined by these embryonic lethal mutations must therefore be transcribed during the blastoderm stage. Molecular hybridization has been used to demonstrate directly that the bithorax complex, which is involved in the determination of segment identity, is newly transcribed at the blastoderm stage (Saint and Hogness, personal communication).

It has been suggested that zygotic genes controlling embryo pattern interpret the embryonic axes established by activity of the maternal genome, and act to subdivide the embryo into smaller units such as segments and tissues (Nusslein-Volhard, 1979; Nusslein-Volhard and Wieschaus, 1980). Transcription from the maternal and zygotic genomes may thus provide qualitatively different kinds of information for pattern formation in the early embryo.

D. Conclusion. During the first few hours of embryogenesis in Drosophila, events occur which are crucial to later development: extremely rapid nuclear multiplication, cellularization and activation of transcription. Of greater interest, perhaps, is the determination of cells of the blastoderm to give rise to the basic body plan of the embryo. Genetic and biochemical evidence indicate that only a minority of the total genes active during embryogenesis are involved in controlling these processes, and that genes specific to the maternal program may make a qualitatively different contribution to early embryogenesis than genes specific to the zygotic program.

Because there appear to be only a relatively small number of them, it should be possible to define genes involved in processes specific to early embryogenesis by molecular cloning of mRNAs whose concentration changes significantly during the first few hours of embryogenesis. Analysis of such cloned genes and their relationship to known mutations is expected to provide information about: 1) functions which are required for the events specific to early embryogenesis and 2) whether these functions are provided for by the maternal and/or the zygotic portions of the embryonic program. We describe below our procedure for cloning a number of genes specific to early embryogenesis and describe our strategy for analyzing the function of such genes.

Experimental Procedures and Results

A. Rationale. In order to obtain cloned DNA sequences encoding RNAs differentially expressed as part of either the maternal or early zygotic program, we screened two recombinant DNA libraries: a cDNA library of embryo mRNA sequences, and a λ phage library of total Drosophila genomic DNA. Each of the types of libraries (genomic and cDNA) has advantages for screening in terms of subsequent manipulations with cloned sequences. Isolation of sequences from genomic libraries leads more rapidly to cloning of genomic DNA around the mRNA coding sequence of interest, while isolation from cDNA libraries allows one to analyze the expression of a single mRNA species without problems caused by linked mRNA sequences or intervening sequences. We intended, by screening two different libraries in two different ways (see below), to maximize our yield of stage-differential sequences.

The original purpose of these screens was to identify mRNAs whose concentration is greater in the blastoderm embryo than in the oocyte (blastoderm-differential). As will become evident below, sequences more concentrated in the oocyte than in the embryo (maternal-differential) were also obtained.

B. Preparation and screening of the cDNA library. We generated 2500 independent cDNA clones from cytoplasmic poly(A)$^+$ RNA of 2.5-3.5 hr (blastoderm) Drosophila embryos of the Oregon R P2 strain. Because there are approximately 6000 mRNAs in the Drosophila egg mRNA population (Hough-Evans et al., 1980), this library only partially represents the sequence complexity of the poly(A)$^+$ RNA present in blastoderm embryos. There is a 90% probability, however, that it contains any mRNA constituting 0.1% or more of the total mRNA population (Clarke and Carbon, 1976).

For construction of the library, the cDNA strand was synthesized with reverse transcriptase in the presence of actinomycin D, after denaturation of the RNA with methyl mercury hydroxide (Buell et al., 1978; Payvar and Schimke, 1979). The single strand was converted into double stranded cDNA with the Klenow fragment of DNA polymerase

I (Padayatty et al., 1981). The double stranded DNA product was
treated with S1 nuclease (Wickens et al., 1978) and tailed at the
free 3'OH groups with dC residues, using terminal transferase
(Roychoudhury et al., 1976). This DNA was annealed with pBR322
(which had been cut in the amp^r gene with PstI and tailed with dG
residues) and then used to transform E. coli strain Hb101. More than
95% of the tet^r transformants were amp^S, as expected if they carry a
recombinant plasmid with an insert in the amp^r gene. Of 10 amp^{S}tet^r
transformants picked at random, nine were found to have Pst
I-excisable inserts greater than 150 nucleotides (NT) in length
(150-600 NT, average length = 300 NT). Only 1.5% of the clones in
this library contain mitochondrial sequences, as determined by colony
hybridization using ^{32}P-nick-translated Drosophila mitochondrial DNA
(prepared by M. Lusby) as probe.

Fifteen thousand colonies of the amplified cDNA library were
screened for the presence of blastoderm-differential sequences, using
the colony hybridization method of Hanahan and Meselson (1980).
Replicate filters were hybridized with ^{32}P-cDNA made to poly(A)$^+$ RNA
from either blastoderm (2.5-3.5 hr) embryos or hand-dissected stage
13/14 oocytes. All hybridizations were carried out in duplicate, and
only those colonies which were positive on duplicate filters
hybridized with one probe and negative on duplicate filters
hybridized with the other probe were considered to be
stage-differential.

Figure 1 shows the results of a representative differential
screen of two duplicate filters. The great majority of colonies
reacted equally with both probes; this is consistent with the
observation that the great majority of mRNAs sequences are present in
both the oocyte and blastoderm embryo (Arthur et al., 1979). Of the
total colonies which were probed, 5% gave a strong signal, an
additional 40% gave weaker but detectable signals, and the remainder
were not detected. Again, these results are consistent with the
structure of the mRNA population as determined by cDNA-mRNA
hybridization (Arthur et al., 1979) and indicate that we can detect
recombinant plasmids containing inserted mRNA sequences of the high
and intermediate abundance classes.

C. Stage-differential cloned cDNAs. From the initial screen,
42 blastoderm-differential and 32 maternal-differential recombinant
plasmids identified. After rescreening with ^{32}P-cDNA to blastoderm
and ovary poly(A)$^+$ RNA, we obtained 17 and 25 blastoderm- and
maternal-differential plasmids respectively (see Table 1). These
were then rescreened with ^{32}P-cDNA to blastoderm and stage 13/14
oocyte poly(A)$^+$ RNA (Table 1). In this final screen, four
blastoderm-differential and nine maternal-differential recombinant
plasmids were obtained (see Fig. 2A and B). DNA blot hybridization
indicated that all four plasmids containing blastoderm-differential
cDNAs were similar or identical to each other, as were all nine

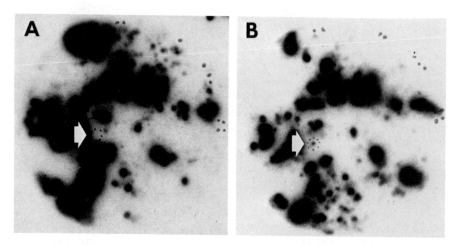

Figure 1: Differential screen of blastoderm cDNA library. ^{32}P-cDNA to maternal (stage 13/14 oocyte) and blastoderm (2.5-3.5 hr embryo) poly(A)$^+$ RNA was used to probe colonies on nitrocellulose filters according to the method of Hanahan and Meselson (1980). Each colony is a transformant from the blastoderm cDNA library (see text). The great majority of colonies react equally with both probes. The white arrows indicate the position (encircled by black dots) of a colony (containing pDO2) which reacts more with maternal ^{32}P cDNA (panel B) than with blastoderm ^{32}P-cDNA (panel A). Although there are other colonies which appear to react differentially in comparing the filters in panels A and B, duplicate experiments showed that these other colonies actually reacted equally with cDNA to RNA of both stages; the colony containing pDO2 was the only one which reacted differentially in a reproducible way on this particular set of filters. The filter shown in panel A is darker because it was exposed somewhat longer in order to detect signal from the pDO2-containing colony with the blastoderm probe.

Table 1. Differential Screen of Blastoderm cDNA Library

Stage of Screen	Stages Compared (^{32}P-cDNA probe)	Source of Colonies Screened	# Stage-differential Plasmids	
			blastoderm	maternal
I. Initial	Blastoderm vs. stage 13/14 oocyte	15,000 colonies (from 2,500 original transformants)	42	32
II. Colony purification	Blastoderm vs. whole ovary	positives from screen I	17	25
III. Rescreen	Blastoderm vs. stage 13/14 oocyte	positives from screen II	4	9

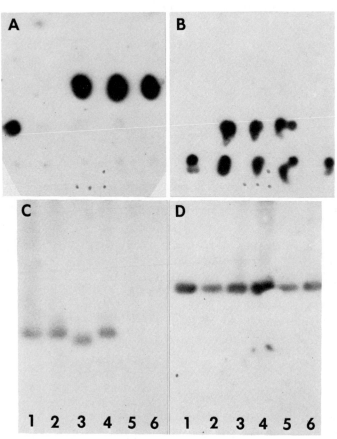

Figure 2: Stage differential recombinant cDNA plasmids: Rescreen
and Characterization. A–B) Twenty-three colonies containing
stage-differential plasmids (from screen 2, described in Table 1)
were patched onto nitrocellulose filters and probed with ^{32}P-cDNA
to blastoderm (A) and stage 13/14 oocyte (B) poly(A)$^+$ RNA. Four
colonies reacted more with the blastoderm than the maternal probe;
these contain the four blastoderm-differential plasmids described
in the text. Eight colonies reacted more with the maternal than
with the blastoderm probe; these contain the maternal-differential
plasmids. Eleven colonies did not react significantly with either
probe and hence were not detected on either autoradiograph. C–D)
DNA from the four blastoderm- and six of the eight
maternal-differential cDNA plasmids from the rescreen shown in
panels A and B were digested with Pst I to excise the insert,
electrophoresed in an agarose gel, blotted onto nitrocellulose
paper, and hybridized with ^{32}P-nick translated DNA of one each of
the blastoderm- (A) and maternal-differential (B) cDNA plasmids.
(C) Blastoderm-differential (lanes 1–4) and two

plasmids containing maternal-differential cDNAs (Fig. 2C and D). The
blastoderm- and maternal-differential cDNA sequences are, however,
clearly distinct from each other (Fig. 2C). The
blastoderm-differential sequence was designated B4 and the maternal
differential sequence O2; the plasmids containing these sequences
were designated pDB4 and pDO2, respectively.

The plasmid pDB4 has an insert size of 300 NT and in an RNA
blot, hybridizes to a poly(A)$^+$ RNA of approximately 1100 NT. This
RNA is most abundant at the late gastrula stage (5-6 hr). The B4
sequence hybridizes in situ to many loci on the Drosophila polytene
chromosomes. These characteristics suggest that the B4 sequence may
be related to the repeated sequence B104 which was isolated as a
blastoderm-differential sequence by Scherer et al. (1981). The B4
sequence has not been studied further.

D. Characteristics of the maternal mRNA sequence O2. The
plasmid pDO2 contains an insert 530 NT in length. Only one poly(A)$^+$
RNA species, 800 NT in length, is homologous to the O2 insert.
Hybridization of ^{32}P-nick-translated pDO2 probe to equal aliquots of
poly(A)$^+$ RNA from different stages of embryogenesis shows that the O2
mRNA is most abundant in newly laid embryos (0-1 hr) and declines in
concentration thereafter with a half life of 1-2 hr (Fig. 3A). This
pattern of temporal expression suggests that the O2 sequence is
transcribed primarily during oogenesis. Consistent with this
interpretation, O2 mRNA is found to be present at a relatively high
level in whole ovaries, but is not detected in the female carcass
minus ovaries (Fig. 3B). O2 mRNA is just barely detectable in whole
male adults and testes (Fig. 3B) and in cultured cells (data not
shown). Quantitative filter hybridization analyses (data not shown)
indicate that the O2 mRNA constitutes slightly more than 0.1% of the
total mRNA in 0-1 hr embryos. Very little O2 RNA is detected in the
poly(A)$^-$ fraction of embryos or ovaries (data not shown). The
characteristics of its expression indicate that O2 is a maternal mRNA
which acts primarily during late oogenesis or early embryogenesis,

Figure 2 (continued):

maternal-differential (lanes 5 and 6) cDNA plasmids probed with
^{32}P-nick translated DNA of one of the blastoderm-differential
plasmids. (D) Six of the maternal-differential plasmids (lanes
1-6) probed with ^{32}P-nick translated DNA of one of the
maternal-differential plasmids. All of the blastoderm sequences
share homology with each other, as well as do all of the maternal
sequences, but there is no homology between the two classes. The
isolation of plasmids containing what appears to be the same cDNA
sequence was not unexpected since the original cDNA library
containing 2,500 transformants was amplified to 15,000 colonies.

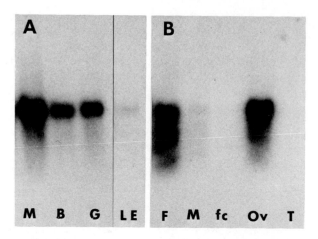

Figure 3: Expression of RNA encoded by the O2 sequence. One µg of poly(A)+ RNA from different stages of embryogenesis, whole adults, and adult tissues was denatured with glyoxal, electrophoresed in a 1.4% agarose gel, blotted onto nitrocellulose paper, and probed with 32P-nick translated DNA of the plasmid pDO2.

A) RNA from different stages of embryogenesis (M = 0.25-1.25 hr; B = 2.5-3.5 hr; G = 5-6 hr; LE = 17-19 hr).

B) RNA from whole adults (F = female; M = male) and adult tissues (Ov = ovary; T = testis; and fc = female carcass without ovaries).

The autoradiograph in panel B was exposed five times as long as that in panel A to allow detection of very small amounts of O2 mRNA in adult males. The concentration of O2 mRNA in ovary poly(A)+ RNA is actually less than 80% of that in 0-1 hr embryo poly(A)+ mRNA.

but which may also be required, at a much lower level, at other stages in the life cycle.

The copy number of the O2 sequence was investigated by the DNA blot reconstruction method of Lis et al. (1979). Amounts of PstI-restricted pDO2 DNA equivalent to four, two, one, and one-half copies of O2 per haploid genome were electrophoresed adjacent to Sst I-restricted total embryonic DNA, blotted onto nitrocellulose, and probed with 32P-nick-translated pDO2 DNA (Fig. 4). The intensity of the 530NT O2 cDNA insert relative to that of the single Sst I genome fragment indicates that the O2 sequence is present in a small number of copies (not more than three) per genome. If O2 is present in multiple copies, these must be clustered in the genome as there is only one 3.2 kb Sst I (Fig. 4) and one 7.1 kb EcoRI genomic fragment which reacts with the pDO2 probe (see below).

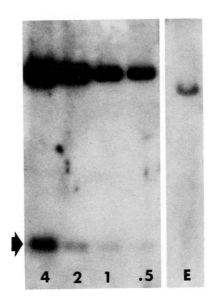

Figure 4: Estimate of O2 copy number by reconstruction method. Five
μg of embryonic DNA was digested with Sst I (lane E) and
electrophoresed adjacent to Pst I-digested pDO2 DNA containing
amounts of insert corresponding to four, two, one, and one-half
copies per haploid genome. After electrophoresis the DNA was
blotted onto nitrocellulose paper and probed with ^{32}P-nick
translated pDO2 DNA. The lower arrow indicates the 530NT O2
insert, the upper arrow the single 3.2 kb Sst I genomic fragment
which hybridizes with the O2 probe. Densitometric scans of the
autoradiograph suggest that the O2 sequence is present in
approximately 2-3 copies per haploid genome.

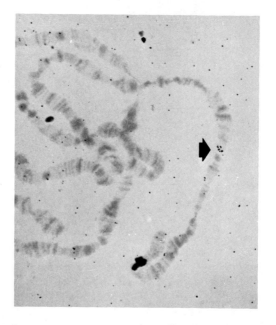

Figure 5: Chromosomal location of O2. ^{3}H-cRNA to the plasmid pDO2
(200,000 cpm/slide) was hybridized in situ to cytological
preparations of Drosophila polytene chromosomes as described by
Lengyel et al. (1980). Exposure 49d.

A

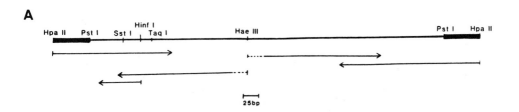

B

```
        10          20          30          40          50          60          70          80          90
···TCC AGG TGT GCA TGG ACG TCG CCC AGG TTC AAG CCC AGT GAG CTC AAC GTG AAG GTG GTG GAC GAC TCC ATC TTG GTC GAG GNC AAG CAT
···ser arg cys ala trp thr ser pro arg phe lys pro ser glu leu asn val lys val val asp asp ser ile leu val glu     lys his

        100         110         120         130         140         150         160         170         180
GAG GAA CGC CAG GAC GAC CAG TGT CAC ATC ATG TCG CCA CTT TGT GCG CCG CTA CAA GGT TCC CGA TGG CTA CAA GGC GGA GCA AGT GGT
glu glu arg gln asp asp gln cys his ile met ser pro leu cys ala pro leu gln gly ser arg trp leu gln gly gly ala ser gly

        190         200         210         220         230         240         250         260         270
CTC GCA GCT GTC GTC GGA TGG CGT GCT CAC NNN NNN NNN NNN NNN NNN NNN NNN NNN NNN NNN NNN NNN NGA GAA TCC AAG GAG CGC ATT
leu ala ala val val gly trp arg ala his                                                     glu ser lys glu arg ile

        280         290         300         310         320         330         340         350         360
CAA ATT CAG CAA GTG GGA CCC GCT CAC CTC AAC GTT AAG GCA AAT GAA AGC GAG GTG AAG GGC AAG GAG AAC GGA GCA CCC AAC GGC AAG
gln ile gln gln val gly pro ala his leu asn val lys ala asn glu ser glu val lys gly lys glu asn gly ala pro asn gly lys

        370         380         390         400         410         420         430         440         450
GAC AAG TAA AGG AGC CAT CAT CAT CCA ACA TCA TCC ATC ATC ATT CCC CTG ACT TAA TTG TTC CTA ATT TAT TGC ATT GTA TTT GTA ATG
asp lys end

        460         470         480         490         500         510         520
AGC TAA AGA CTA GAA TAC TCA TAT TAA TTT AAT AAA TCC TTT TGT TCA ACC TGG TGT GGA AAA AAA AAA AAA AAA AAA
```

Figure 6:   Sequencing strategy and partial sequence of 02 cDNA.

A) Restriction map and sequencing strategy. The thin line
represents the 02 cDNA inserted into the Pst I site of pBR322
(flanking sequences of which are depicted by heavy lines). The
arrows represent 5'-end labeled fragments used for DNA sequencing
(Maxam and Gilbert, 1980). A small portion of the sequence has not
yet been obtained due to technical difficulties encountered in
sequencing a fragment extending through the Hae III site. The
direction of transcription is from left to right (see text).

B) Partial sequence of 02 cDNA. The nucleotide sequence (upper
case) is shown above a proposed amino acid sequence (lower case)
(see text for explanation). N represents nucleotides for which the
sequence is uncertain. Note the several stop codons (underlined),
the poly(A) addition signal (bold underline), and the poly(A) tract
at the end of the sequence.

In situ hybridization analysis indicates that the 02 sequence is located in only one region in the polytene chromosome, 67B (Fig. 5). Since the 02 sequence does not hybridize to any other chromosomal locus, and hybridizes to only a single Sst I and EcoRI fragment, all copies of 02 (if there are more than one) must be at this chromosomal locus.

In order to gain insight into the function of the 02 gene, a partial nucleotide sequence of the 02 cDNA has been determined using the base-specific chemical cleavage method of Maxam and Gilbert (1980) and the strategy shown in Fig. 6A. The size of the cDNA insert by restriction analysis (data not shown) is 530+/-10 NT; of this, a partial sequence of 488 bp has been obtained (Fig. 6B). The presence of a poly(A) addition signal (Proudfoot and Brownlee, 1976; Benoist et al., 1979; Torok and Karch, 1980) and a terminal poly(A) tract are evidence that this cDNA insert represents the 3' end of the 02 mRNA. This is corroborated by an experiment which showed that 02 mRNA hybridized only the sequence complementary to the one shown in Fig. 6B (data not shown).

A computer search (Queen and Korn, 1980) was performed to identify open reading frames in the 02 cDNA sequence. Using the direction of transcription shown, only one open reading frame, translated below the nucleotide sequence into a proposed amino acid sequence (Fig. 6B), is found which includes the first 210 nucleotides. For the final 276 nucleotides, the proposed amino acid sequence is that dictated by the longest open reading frame. The proposed amino acid sequence shown represents more than half of the residues encoded by the 800 NT 02 mRNA. In the proposed partial amino acid sequence there is approximately equal representation of positively and negatively charged residues (15 vs. 14) and a somewhat greater number of polar than nonpolar residues (57 vs. 44). When a complete nucleotide sequence of 02 has been obtained, a search for homology with known sequenced genes and proteins will be undertaken to suggest possible functions of the 02 gene product.

E. Isolation of DNA around 02. The isolation of genomic DNA containing the 02 sequence and flanking it on both sides should facilitate identification of sequences which might be involved in 02 regulation, as well as construction of a physical map of the region in which the 02 sequence can be localized relative to deletion endpoints and "point" mutations within the region.

Genomic DNA homologous to the 02 cDNA sequence was isolated by screening (Benton and Davis, 1977) a partially EcoRI-restricted

Drosophila genomic library in λ Charon 4 (Yen et al., 1979) with the Pst I-excised 530 NT cDNA sequence which had been nick-translated with ^{32}P. All of the genomic DNA from the phage isolated in this screen overlapped and shared a common 7.1 kb EcoRI restriction fragment (Fig. 7). These results are consistent with the in situ hybridization of the O2 sequence to only one chromosomal locus (Fig. 5) with the estimate of copy number showing three or fewer O2 sequences present on a single Sst I fragment (Fig. 4) and with DNA blot analysis of EcoRI-restricted adult and embryo DNA showing hybridization of O2 to a single 7.1 kb EcoRI fragment (data not shown). By rescreening the genomic library using the most distal fragments obtained as probes, a series of overlapping genome DNA fragments representing a total of 45 kb of DNA around O2 has been isolated (Fig. 7). The overlap hybridization screen is being continued.

F. Isolation of blastoderm- and gastrula-differential sequences. As an alternative means of obtaining stage-differential genomic sequences, we screened a Drosophila genomic library in λ phage as described above. Because cDNA probes might not be representative of all mRNAs in the population (Scherer et al., 1981), the library was screened with base cleaved, ^{32}P-kinased poly(A)$^+$ RNA isolated from polysomes of blastoderm embryos. The high sequence complexity of embryo mRNA means that a large number of phage are expected to react with both maternal and blastoderm probes (see, for example, Fig. 1); this would make it difficult to detect differential expression of a phage encoding mRNAs of low to intermediate concentration. For this reason, a competition screen (Mangiarotti, et al., 1981) was carried out: the ^{32}P-kinased blastoderm poly(A)$^+$ RNA probe was hybridized in the presence of a 35-fold excess of unlabeled poly(A)$^+$ cytoplasmic RNA from 0.5-1.5 hr embryos.

Twenty-five thousand phage were screened in this way, giving a 90% chance of detecting a single copy gene (Clarke and Carbon, 1976). All of the phage giving the most intense signals on duplicate filters, most giving moderately intense signals, and a random selection from the large number giving a pale signal, were selected in this screen. These 266 phage were plaque purified and differentially screened twice using maternal (0.5-1.5 hr embryo) and blastoderm (2.5-3.5 hr embryo) ^{32}P-cDNA. A total of 15 blastoderm- and 35 maternal-differential phage, representing 5.6 and 13%, respectively, of the total plaques selected in the initial screen, were obtained.

A number of conclusions can be drawn from these results. The phage encoding mRNAs whose concentration changes during the first 3.5 hr of Drosophila embryogenesis constitute a small fraction of the total phage library. This is consistent with the cDNA-mRNA hybridization experiments, with analyses of protein synthesis and with genetic studies, and indicates that only a small number of mRNAs change in concentration during early embryogenesis. For those mRNAs whose concentrations do change significantly, more maternal-differential than blastoderm-differential sequences were obtained by the screening procedures used.

G. Characteristics of some blastoderm- and gastrula-differential sequences. The expression of nine of the 15 blastoderm-differential genomic sequences has been characterized by RNA blot hybridization. Of these partially characterized recombinant phage, six have been found to contain genomic sequences encoding mRNAs which are either blastoderm- or late gastrula-differential (Fig. 8).

These phage contain genomic inserts of 13-21 kb (Table 2); most encode only one mRNA. Messenger RNAs encoded by the various phage appear, by their electrophoretic migration relative to Drosophila 19S rRNA, to differ in size. Only one genomic EcoRI fragment from most of these phage hybridizes with ^{32}P-cDNA from synthesized blastoderm poly(A)$^{+}$ RNA; in each case the size of the hybridizing fragment is different (Table 2, third column).

Two phage encode more than one RNA. These are 1531, which encodes a minimum of two mRNAs, at least one of which is expressed differentially at the blastoderm stage (Fig. 8), and 1561, which encodes three mRNA species, one of which is differentially expressed at the blastoderm stage (Table 2). Both of these phage carry two or more restriction fragments which hybridize to blastoderm cDNA (Table 2). These results, together with the difference in developmental regulation among the sequences encoded by each phage, suggest that there is more than one transcription unit on each of phage 1531 and 1561. Further restriction and hybridization analysis will be necessary to confirm this point.

The mRNAs encoded by the various phage appear to be present in widely different concentrations. The most abundant RNA, that encoded

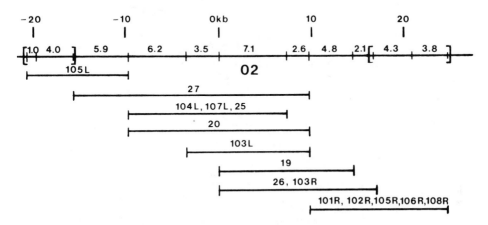

Figure 7: Overlapping genomic DNA fragments around 02. A set of
overlapping genomic DNA sequences in phage λ were obtained by
screening a partial EcoRI λ library (Yen et al., 1979) with
^{32}P-nick translated 02 sequence (excised from pDO2 with Pst I).
From this screen the phage 27, 25, 20, 19, and 26 were obtained.
The 5.9 kb fragment from phage 27 and the 2.1 kb fragment from
phage 26 were then purified by electrophoresis in agarose, nick
translated with ^{32}P, and used to rescreen the library. From this
screen the phage 105L, 104L, 107L, and 101, 102, 103, 105, 106,
108 R were obtained. The vertical lines represent EcoRI sites.
The EcoRI fragments in brackets have not been ordered relative to
each other because no phage have been obtained which contain one
fragment but not the other. The "walk" is being continued using
the fragments in brackets at either side.

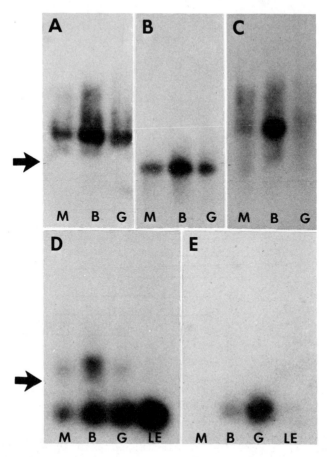

Figure 8:   Changing expression of mRNAs homologous to blastoderm- and
gastrula-differential genomic sequences. Two µg of poly(A)$^+$ mRNA
from different stages of embryogenesis (M = 0.5-1.5 hr; B = 2.5-3.5
hr; G = 5-6 hr; LE = 17-18 hr) were denatured with glyoxal and
electrophoresed in agarose gels. RNA in the gels was blotted onto
nitrocellulose paper and hybridized to ^{32}p-nick translated DNA
(0.08-5 x 10^8 dpm/µg) of various λ phage which had been selected in
the differential screen (see text). The arrows indicate the
migration in the gel of a 2000-2300 NT marker (Drosophila 19S and
denatured 26S rRNA). Exposure times were selected to be within the
linear range of the film and varied widely for the different phage
probes (from 2 hr for 542 to 168 hr for 1351). A = phage 361; B =
phage 542; C = phage 1503; D = phage 1531; E = phage 1351.

Table 2.   Characterization of Recombinant Phase Containing
Blastoderm- and Gastrula Differential Sequences

Phage #	Message Species[a] total/differential	Relative[b] Concentration at Blastoderm	EcoRI Restriction Fragments (kb)[c]	Chromosomal Location[d]
361	1+ / 1+	++	n.d.	99EF, 55A–F chromo-center, nucleolus
542	1 / 1	+++	5.3*, 11.5	99EF, nucleolus, chromo-center
1351	1 / 1	+	9.2,* 4.0	71A
1503	1 / 1	+	9.4, 7.0*	25D
1531	2+ / 1+	+	7.8, 5.9,* 5.0,* 2.6	99DE
1561	3 / 1	+	7.5,* 4.0,* 2.8*	94F/95A

[a]Determined by gel electrophoresis of embryo poly(A)+ RNA, probed with ^{32}P-nick translated DNA from each recombinant λ phage, as shown in Fig. 8. In some cases (indicated by +) there were suggestions of additional faint mRNA species in the autoradiograms (for example Fig. 8)

[b]A qualitative estimate, determined by scanning the blastoderm lanes of gel autoradiograms (Fig. 8 and data not shown) and normalizing to the amount of RNA in the lane, to the hours of exposure of the film and to the specific activity of the probe.

[c]Each phage was digested completely with EcoRI and electrophoresed in agarose. The sizes of the inserted Drosophila genomic fragments are relative to Hind III-digested λ. The fragments in the gel were blotted onto nitrocellulose paper and probed with ^{32}P-cDNA to blastoderm poly(A)+ RNA. The asterisks (*) indicate the phage fragments which hybridized with the ^{32}P-cDNA probe.

[d]Determined by in situ hybridization of ^{3}H-cRNA from each phage to polytene chromosomes, as in Fig. 5.

by 5402, appears to be more abundant than actin mRNA, which we have shown previously to constitute 0.15% of the total poly(A)$^+$ mRNA at the blastoderm stage (Anderson and Lengyel, 1983). The mRNAs encoded by the other phage shown in Table 2, with the exception of 361, are all present at a much lower concentration, possibly 100-fold less than the 542 mRNA.

The mRNA species encoded by one of the phage (1351) is first detected at blastoderm, increases substantially by late gastrula, and then decays to an undetectable level in late embryos (Fig. 8). A number of other phage isolated in this screen for blastoderm-differential sequences were also found to contain genomic sequences encoding mRNAs maximally expressed at the late gastrula stage (data not shown).

The blastoderm- and gastrula-differential genomic sequences in the recombinant phage shown in Table 2 hybridize to a number of different chromosomal loci. The genomic sequences in phage 1351, 1503, 1561 and 1531 are present at four different single chromosomal loci. The genomic sequence in phage 542 hybridizes to one euchromatic site, but also to the nucleolus and chromocenter. Finally, the genomic DNA of phage 361 hybridizes to two euchromatic sites, a major and a minor site, as well as to the nucleolus and chromocenter. This hybridization to the nucleolus and chromocenter is probably due to the presence, in these latter two phage, of repeated genomic sequences. It is noteworthy that three of the phage, 361, 542, and 1531, have similar chromosomal location sites (99D-F). While the size and abundance of the blastoderm-differential mRNAs encoded by these phage would suggest that they are different sequences, we cannot rule out the possibility that the genomic DNAs in the three phage are overlapping. Experiments to test this possibility are in progress.

Although the results of the analysis described above are very preliminary, it is possible to come to a number of conclusions. It appears that there are a number of different sequences which are differentially expressed at the blastoderm and gastrula stages. We have not found any sequences which appear to be absolutely blastoderm-specific; this may be due to a low level asynchrony in the staged embryos used for the mRNA preparation. It may be that absolutely stage-specific mRNA sequences are very rare.

DISCUSSION

The initial hypothesis of this research, based upon both hybridization studies and genetic analyses, is that there are genes whose expression changes dramatically during early embryogenesis, that such genes are small in number, and that such genes can be cloned by differential or competition hybridization. The results we have obtained here are consistent with this hypothesis. We have obtained cDNA and genomic sequences coding for two types of mRNAs differentially expressed in early Drosophila embryogenesis. One type of mRNA is part of the maternal mRNA population; immediately after fertilization this mRNA decays so that it is greatly reduced in concentration by the end of embryogenesis. Another class of RNA is differentially expressed in embryogenesis: this class is present in low or undetectable amounts in the maternal mRNA population (0-1 hr embryos), increases to reach a maximum at blastoderm (2.5-3.5 hr) or late gastrula (5-6 hr), and then decreases to undetectable amounts in the late embryo (17-19 hr).

Other laboratories have carried out studies to identify mRNA sequences which are differentially expressed early in Drosophila embryogenesis. Fifteen maternal-differential sequences have been identified by E. Stephenson and A. Mahowald (personal communication); none of these appear, on the basis of chromosomal mapping, to be related to the 02 sequence described here. Five blastoderm and gastrula-differential sequences were identified by Scherer et al., (1981); of the three for which chromosomal localizations were obtained, none overlapped with the sequences obtained in the work described here. Another study identified several sequences which are differentially expressed early in Drosophila embryogenesis (Sina and Pellegrini, 1982); because chromosome locations were not obtained for these sequences, we do not know if they are related to those obtained in this study. We conclude that the total number of blastoderm and gastrula-differential sequences present in the Drosophila embryo, although small compared to the total number of mRNAs in the population, has not yet been obtained by molecular cloning.

The other portion of the hypothesis to be tested in this work is that differentially expressed sequences are involved in controlling processes important to early embryogenesis. Analysis of the function of a number of such sequences is likely to provide additional information about biochemical processes already known to occur in early embryogenesis. Such analysis may also provide information about functions involved in determination, about which very little is known at the biochemical level. Analysis of cloned genes which are maternal-differential and zygotic-differential might give further insight into the nature of the programs which are established as a result of maternal transcription and those which are established, epigenetically, by zygotic transcription.

The function of sequences which have been cloned can be approached by both genetic and biochemical methods. Some of the blastoderm-differential sequences may map near genes known to affect the cell determination and pattern formation which occurs at blastoderm. This can be determined by comparison of the chromosomal location of cloned sequences, determined by in situ hybridization, with the location of 120 embryonic lethal mutations which affect embryonic pattern, determined by recombination and cytology (Nusslein-Volhard et al., 1983; Nusslein-Volhard, personal communication). Whether a cloned sequence actually corresponds to a known genetic locus can only be determined by DNA blot analysis; this requires the existence either of mutant alleles which consist of DNA rearrangements, or of closely mapped DNA rearrangements or deletions which define the locus. The correlation of the genetic with the DNA map has been achieved for the whit and scute loci (Levis et al., 1982; Carramolino et al., 1982).

An alternative approach to understanding gene function is the biochemical characterization of the gene product. The polypeptide sequence predicted from the nucleotide sequence of the mRNA can be compared for possible homologies to the 2500 catalogued protein sequences. An additional powerful approach is to synthesize a small portion (10-15 amino acids) of the predicted polypeptide and make antibodies to this small peptide (Lerner et al., 1981; Lerner, 1982). These antibodies can then be used to analyze the temporal and spatial distribution of the gene product during development. This technique has been applied successfully to identify a previously unknown protein predicted by a viral genomic sequence (reviewed by Lerner et al., 1981; Lerner, 1982).

We conclude that it is possible to obtain, by genomic or cDNA cloning, sequences encoding mRNAs whose concentration changes significantly during early embryogenesis. Some of these mRNAs are relatively rare (probably 0.01% or less of the total mRNA population). Such RNAs are likely to play a role in controlling events of early embryogenesis. By investigating the function of such cloned sequences by both genetic and biochemical analyses, we should increase our understanding of the unique processes occurring during early embryogenesis.

ACKNOWLEDGEMENTS

    This work was supported by grants from the NIH (HD-09948) and NSF (PCM 21830) to JAL, by an NIH fellowship (GM 08809) to EMU and by training grants USPHS GM 07185 to PDB and GM 07104 to MR. We are grateful to our colleagues Drs. Allan J. Tobin and John R. Merriam for helpful comments on the manuscript.

REFERENCES

Anderson, K.V. and Lengyel, J.A. (1979). Rates of synthesis of major classes of RNA in Drosophila embryos. Devel. Biol. 70:217-231.

Anderson, K.V. and Lengyel, J.A. (1981). Changing rates of DNA and RNA synthesis in Drosophila embryos. Devel. Biol. 82:127-138.

Anderson, K.V. and Lengyel, J.A. (1983). Histone gene expression in Drosophila development: Multiple levels of gene regulation. In "Histone Genes and Histone Gene Expression," G. Stein, J. Stein, and W. Marzluff (eds.). John Wiley and Sons, in press.

Anderson, K.V. and Nusslein-Volhard, C. (1983). Genetic analysis of dorsal-ventral embryonic pattern in Drosophila. In "Drosophila in Pattern Formation," G. Malacinski and S. Bryant (eds.), in press.

Arthur, C.G., Weide, C.M., Vincent, W.S., and Goldstein, E.S. (1979). mRNA sequence diversity during early embryogenesis in Drosophila melanogaster. Exp. Cell. Res. 121:87-94.

Bakken, A.H. (1973). A cytological and genetic study of oogenesis in Drosophila melanogaster. Devel. Biol. 33:100-122.

Benoist, C., O'Hare, K.O., Breathnach, R., and Chambon, P. (1979). The ovalbumin gene-sequence of putative control regions. Nuc. Acadis Res. 8:127-142.

Benton, W.D. and Davis, R.W. (1977). Screening recombinant clones by hybridization to single plaques in situ. Science 196:180-182.

Buell, G.N., Wickens, M.P., Payvar, F., and Schmike, R.T. (1978). Synthesis of full-length cDNA from four partially purified oviduct mRNAs. J. Biol. Chem. 253:2471-2482.

Capdevila, M.P. and Garcia-Bellido, A. (1974). Development and genetic analysis of bithorax phenocopies in Drosophila. Nature 250:500-502.

Carramolino, L., Ruiz-Gomez, M., Guerrero, M.d.C., Campuzano, S., and Modolell, J. (1982). "DNA map of mutations at the scute locus of Drosophila melanogaster." The EMBO Journal I:1185-1191.

Chan, L-N. and Gehring, W. (1971). Determination of blastoderm cells in Drosophila melanogaster. Proc. Natl. Acad. Sci. USA 68:2217-2221.

Clarke, L. and Carbon, J. (1976). A colony bank containing synthetic ColEl hybrid plasmids representative of the entire E. coli genome. Cell 9:91-99.

Cline, T.W. (1976). A sex-specific, temperature-sensitive maternal effect of the daughterless mutation of Drosophila melanogaster. Genetics 84:723-742.

Cline, T.W. (1978). Two closely linked mutations in Drosophila melanogaster that are lethal to opposite sexes and interact with daughterless. Genetics 90:683-698.

Fullilove, S.L. and Jacobson, A.G. (1978). Embryonic development: descriptive. In "The Genetics and Biology of Drosophila," Vol. 2c, M. Ashburner and T.R.F. Wright (eds.), pp. 105-227. Academic Press, New York.

Gans, M., Audit, C., and Masson, M. (1975). Isolation and characterzation of sex-linked female sterile mutants in Drosophila melanogaster. Genetics 81:683-704.

Goldstein, E.S. and Arthur, C.G. (1979). Isolation and characterization of cDNA complementary to transient maternal poly(A)$^+$ RNA from the Drosophila oocyte. Biochim. Biophys. Acta 565:265-274.

Goldstein, E.S. and Snyder, L.A. (1973). Protein synthesis and distribution of ribosomal elements in ovarian oocytes and developmental stages of Drosophila melanogaster. Exp. Cell Res. 81:47-56.

Gutzeit, H.O. (1980). Expression of the zygotic genome in
    blastoderm stage embryos of Drosophila: an analysis of a
    specific protein. Wilhelm Roux' Archiv. 188:153-156.

Hanahan, D. and Meselson, M. (1980). Plasmid screening at high
    colony density. Gene 10:63-67.

Hough-Evans, B.R., Jacobs-Lorena, M., Cummings, M.R., Britten, R.J.,
    and Davidson, E.H. (1980). Complexity of RNA in eggs of
    Drosophila melanogaster and Musca domestica. Genetics
    95:81-94.

Illmensee, K. and Mahowald, A.P. (1974). Transplantation of
    posterior polar plasm in Drosophila: Induction of germ cells at
    the anterior pole of the egg. Proc. Natl. Acad. Sci. USA
    71:1016-1020.

Illmensee, K. and Mahowald, A.P. (1976). The autonomous function of
    germ plasm in a somatic region of the Drosophila egg. Exp.
    Cell. Res. 97:127-140.

Illmensee, K., Mahowald, A.P., and Loomis, M.R. (1976). The
    ontogeny of germ plasm during oogenesis in Drosophila. Devel.
    Biol. 49:40-65.

Jackle, H. and Kalthoff, K. (1979). RNA and Protein Synthesis in
    Developing Embryos of Smittia spec. (Chironomidae, Diptera).
    Wilhelm Roux' Archiv. 187:283-305.

Janning, W. (1978). Gynandromorph fate maps in Drosophila. In
    "Genetic Mosaics and Cell Differentiation," W.J. Gehring (ed.),
    pp. 1-28. Springer Verlag, New York.

Kalthoff, K. (1979). Analysis of a morphogenetic determinant in an
    insect embryo. In "Determinants of Spatial Organization," I.
    Konigsberg and S. Subtelney (eds.), pp. 97-126. Academic Press,
    New York.

King, R.C. and Mohler, D.J. (1975). The genetic analysis of
    oogenesis in Drosophila melanogaster. In "Handbook of
    Genetics," Vol. 3, R.C. King (ed.), pp. 757-791. Plenum
    Publishing, New York.

Konrad, K.D., Goralski, T.J., Turner, F.R., and Mahowald, A.P.
    (1982). Maternal effect mutation affecting gastrulation. J.
    Cell Biol. 95:159a.

Lerner, R.A. (1982). Tapping the immunological repertoire to
    produce antibodies of predetermined specificity. Nature
    299:592-596.

Lerner, R.A., Sutcliffe, J.G., and Shinnick, T.M. (1981). Antibodies to chemically synthesized peptides predicted from DNA sequences as probes of gene expression. Cell 23:309-310.

Levis, R., Bingham, P.M., and Rubin, G.M. (1982). Physical map of the white locus of D. melanogaster. Proc. Natl. Acad. Sci. USA 79:564-568.

Lindsley, D.E. (1979). "Some developmental effects of three mutations which map within a tighly linked cluster of maternal-effect genes." M.S. Thesis, University of Washington.

Lis, J.T., Prestidge, L., and Hogness, D.S. (1979). A novel arrangement of tandemly repeated genes at a major heat shock site in D. melanogaster. Cell 14:910-920.

Lohs-Schardin, M. (1982). Dicephalic--A Drosophila mutant affecting polarity in follicle cell organization and embryonic patterning. Wilhelm Roux' Archiv. 191:28-36.

Lohs-Schardin, M., Sander, K., Cremer, C., Cremer, T., and Zorn, C. (1979a). Localized ultraviolet laser microbeam irradiation of early Drosophila embryos: fate maps based on location and frequency of adult defects. Devel. Biol. 68:533-545.

Lohs-Schardin, M. Cremer, C., and Nusslein-Volhard, C. (1979b). A fate map for the larval epidermis of Drosophila melanogaster: Localized cuticle defects following irradiation of the blastoderm with an ultraviolet laser microbeam. Devel. Biol. 73:239-255.

Lovett, J.A. and Goldstein, E.S. (1977). The cytoplasmic distribution and characterization of poly(A)$^+$ RNA in oocytes and embryos of Drosophila. Devel. Biol. 61:70-78.

Madhavan, M.M. and Schneiderman, H.A. (1977). Histological analysis of the dynamics of growth of imaginal discs and histoblast nests during the larval development of Drosophila melanogaster. Wilhelm Roux' Archiv. 183:269-305.

Mahowald, A.P. (1971). Polar granules of Drosophila. III. The continuity of polar granules during the life cycle of Drosophila. J. Exp. Zool. 176:329-343.

Mange, A.P. and Sandler, L. (1973). A note on the maternal effect mutants daughterless and abnormal oocyte in Drosophila melanogaster. Genetics 73:73-86.

Mangiarotti, G., Chung, S., Zuker, C., and Lodish, H. (1981).
    Selection and analysis of cloned developmentally regulated
    Dictyostelium discoideum genes by hybridization competition.
    Nuc. Acids Res. 9:947-963.

Maxam, A.M. and Gilbert, W. (1980). Sequencing end-labeled DNA with
    base-specific chemical cleavages. Meth. Enzymol. 65:499-559.

McKnight, S.L. and Miller, Jr., O.L. (1976). Ultrastructural
    patterns of RNA synthesis during early embryogenesis of
    Drosophila melanogaster. Cell 8:305-319.

Nusslein-Volhard, C. (1979). Maternal effect mutations that alter
    the spatial coordinates of the embryo. In "Determinants of
    Spatial Organization," S. Subtelney (ed.), pp. 185-211.
    Academic Press, New York.

Nusslein-Volhard, C. and Wieschaus, E. (1980). Mutations affecting
    segment number and polarity in _Drosophila_. Nature 287:795-801.

Nusslein-Volhard, C., Wieschaus, E., and Kluding, H. (1983). Lethal
    mutations affecting the pattern of the larval cuticle in
    Drosophila melanogaster. Wilhelm Roux' Archiv., in press.

Okada, M., Kleinman, I.A., and Schneiderman, H.A. (1974).
    Restoration of fertility in sterilized _Drosophila_ eggs by
    transplantation of polar cytoplasm. Devel. Biol. 37:43-54.

Padayatty, J., Cummings, I., Manske, C.L., Higuchi, R., Woo, S., and
    Salser, W. (1981). Cloning of chicken globin cDNA in bacterial
    plasmids. Gene 13:417-422.

Payvar, F. and Schimke, R.T. (1979). Methylmercury hydroxide
    enhancement of translation and transcription of ovalbumin and
    conalbumin mRNAs. J. Biol. Chem. 254:7636-7642.

Poulsen, D.F. (1950). Histogenesis, organogenesis and
    differentiation in the embryo of _Drosophila melanogaster_ Meigen.
    In "Biology of _Drosophila_," M. Demerec (ed.). Hafner Publishing
    Co., New York.

Proudfoot, N.J. and Brownlee, G.G. (1976). 3' non-coding region
    sequences in eukaryotic messenger RNA. Nature 263:211-214.

Queen, C. and Korn, L. (1980). Computer analysis of nucleic acids
    and proteins. Meth. Enzymol. 65:595-609.

Rabinowitz, M. (1941). Studies on the cytology and early embryology
    of the egg of _Drosophila melanogaster_. J. Morphol. 69:1-49.

Rice, T.B. and Garen, A. (1975). Localized defects of blastoderm formation in maternal effect mutants of Drosophila. Devel. Biol. 43:277-286.

Roychoudhury, R., Jay, E., and Wu, R. (1976). Terminal labeling and addition of homopolymer tracts to duplex DNA fragments by terminal deoxynucleotidyl transferase. Nuc. Acids Res. 3:863-877.

Sakoyama, Y. and Okubo, S. (1981). Two-dimensional gel patterns of protein species during development of Drosophila embryos. Devel. Biol. 81:361-365.

Sandler, L. (1972). On the genetic control of genes located in the sex-chromosome heterochromatin of Drosophila melanogaster. Genetics 70:261-274.

Sandler, L. (1977). Evidence for a set of closely linked autosomal genes that interact with sex-chromosome heterochromatin in Drosophila melanogaster. Genetics 86:567-582.

Santon, J.B. and Pellegrini, M. (1981). Rates of ribosomal protein and total protein synthesis during Drosophila early embryogenesis. Devel. Biol. 85:252-257.

Savoini, A. Micali, F., Marzari, R., de Cristini, F., and Grazioski, G. (1981). Low variablity of the protein species synthesized by Drosophila melanogaster embryos. Wilhelm Roux' Archiv. 190:161-167.

Scherer, G., Telford, J., Baldani, C., and Pirrotta, V. (1981). Isolation of cloned genes differentially expressed at early and late stages of Drosophila embryonic development. Devel. Biol. 86:438-447.

Simcox, A.A. and Sang, J.H. (1983). When does determination occur in Drosophila embryos? Devel. Biol., in press.

Sina, B.J. and Pellegrini, M. (1982). Genomic clones coding for some initial genes expressed during Drosophila development. Proc. Natl. Acad. Sci. USA 79:7351-7355.

Sonnenblick, B.P. (1950). The early embryology of Drosophila melanogaster. In "Biology of Drosophila," M. Denerec (ed.). Hafner Publishing Co., New York.

Struhl, G. and Brower, D. (1982). Early role of the esc[+] gene product in the determination of segments in Drosophila. Cell 31:285-292.

Swanson, M.M. and Poodry, C.A. (1980). Pole cell formation in Drosophila melanogaster. Devel. Biol. 75:419-430.

Thierry-Mieg, D. (1976). Study of a temperature-sensitive mutant grandchildless-like in Drosophila melanogaster. J. Microsc. Biol. Cell. 25:1-6.

Thomas, P.S. (1980). Hybridization of denatured RNA and small DNA fragments transferred to nitrocellulose. Proc. Natl. Acad. Sci. USA 77:5201-5205.

Torok, I. and Karch, F. (1980). Nucleotide sequences of heat shock activated genes in Drosophila melanogaster. I. Sequences in the regions of the 5' and 3' end of the hsp70 gene in the hybrid plasmid 56H8. Nucl. Acids. Res. 8:3105-3123.

Underwood, E.M., Turner, F.R., and Mahowald, A.P. (1980). Analysis of cell movements and fate mapping during early embryogenesis in Drosophila melanogaster. Devel. Biol. 74:286-301.

Warn, R.M. and Magrath, R. (1982). Observations by a novel method of surface changes during the syncytial blastoderm stages of the Drosophila embryos. Devel. Biol. 89:540-548.

Wickens, M.P., Buell, G.N., and Schimke, R.T. (1978). Synthesis of double-stranded DNA complementary to lysozyme, ovomucoid and ovalbumin mRNAs. J. Biol. Chem. 253:2483-2495.

Wieschaus, E. and Gehring, W. (1976). Clonal analysis of primordial disk cells in the early embryo of Drosophila melanogaster. Devel. Biol. 50:249-263.

Wright, T.R.F. (1970). The genetics of embryogenesis of Drosophila. Advan. Genet. 15:261-395.

Yen, P., Hershey, N.D., Robinson, R., and Davidson, N. (1979). Sequence organization of Drosophila tRNA genes. In "INC-UCLA Symposium on Eucaryotic Gene Regulation," R. Axel and T. Maniatis (eds.). Academic Press, New York.

Zalokar, M. (1976). Autoradiographic study of protein and RNA formation during early development of Drosophila eggs. Devel. Biol. 49:425-437.

Zalokar, M. Audit, C., and Erk, I. (1975). Developmental defects of female-sterile mutants of Drosophila melanogaster. Devel. Biol. 47:419-432.

Zalokar, M. and Erk, I. (1976). Division and migration of nuclei during early embryogenesis of Drosophila melanogaster. J. Microsc. Biol. Cell. 25:97-106.

ACCUMULATION AND BEHAVIOR OF mRNA DURING OOGENESIS

AND EARLY EMBRYOGENESIS OF XENOPUS LAEVIS

Linda E. Hyman, Hildur V. Colot, and Michael Rosbash[*]

ABSTRACT

Studies examining the messenger RNA population present in Xenopus laevis oocytes and early embryos are discussed. There may be two classes of mRNA present in the oocyte, one which is used by the developing germ cell and the other which is synthesized during early oogenesis and stored for translation after fertilization. During oocyte maturation the mRNA population undergoes a dramatic qualitative change in which the length of the pA tail is altered. The behavior of individual sequences, including actin and ribosomal proteins, is described during maturation and early embryogenesis.

INTRODUCTION

The South African clawed toad, Xenopus laevis, has long been a favorite experimental organism with which to study early development. Germ cell development, fertilization, and embryogenesis have been analyzed using a variety of experimental approaches, ranging from classical, direct manipulations of eggs and embryos to modern techniques of molecular biology. The Xenopus oocyte is particularly well-suited for biochemical investigations because of its unique characteristics; the stage 6 oocyte is an extremely large cell (approximately 1 mm in diameter) with a stored pool of ribosomes and other constituents equivalent to approximately 10^6 average somatic cells. Although normally induced upon fertilization, the early

[*]From the Department of Biology, Brandeis University, Waltham, Massachusetts 02254.

253

phases of embryonic development can be initiated artificially by the investigator. In the presence of Actinomycin D, the fertilized egg is capable of development to approximately the 4,000 cell blastula stage. Since this drug specifically inhibits transcription, these results suggest that gene expression during very early development is exclusively dependent on maternally derived transcription products already present in the oocyte. Both of these classical experiments demonstrate that the unfertilized egg contains most, if not all, of the gene products necessary for development up until the early blastula stage (see Davidson, 1976, for a review).

For this reason, the development of the oocyte itself has also been of considerable interest. In an attempt to learn about gene activity during oogenesis, many investigators have looked for changes in the messenger RNA (mRNA) population during oogenesis as well as upon fertilization and subsequent embryogenesis. In this report, we review some of these data and present new results which may help to elucidate some of the molecular mechanisms responsible for changes in gene activity during these developmental periods.

The ovary of a healthy Xenopus female contains oocytes at all stages of development. Dumont (1972) classified these using morphological criteria, such as size, yolk uptake, and pigmentation. It is relatively easy to prepare large amounts of reasonably pure populations of staged oocytes. In this report, only the early vitellogenic (Dumont stage 2), vitellogenic-maximal lampbrush stage (Dumont stage 3) and mature oocytes (Dumont stage 5) will be discussed. Much work has been done using stage 3 oocytes because of their maximally extended lampbrush chromosomes (Sommerville, 1977). Since these specialized chromosomal structures are extremely active transcriptionally it would be interesting to understand the nature of their transcription products. The classical hypothesis is that these products are the maternal mRNAs which are synthesized and stored during oogenesis and translated during early embryonic development (Davidson, 1976). It was therefore interesting to examine, both quantitatively and qualitatively, the RNA population present in oocytes before and after the lampbrush stage (before and after vitellogenesis). The first studies of this kind measured, by hybridization with ^{3}H-poly(U), the total amount of poly(A)$^+$ RNA present per oocyte as a function of developmental stage (Rosbash and Ford, 1974; Dolecki and Smith, 1979). The accumulation of poly(A)$^+$ RNA is nearly complete before lampbrush chromosomes are maximally extended. The data suggest that most of the poly(A)$^+$ RNA present in the mature oocyte is synthesized in immature cells.

Because vitellogenesis normally occurs over a period of weeks, if not months (Keem et al., 1979), the interpretation stated above deserves further consideration. The significance of a constant amount of poly(A)$^+$ mRNA during vitellogenesis can be explained in at least four ways: 1) developmental changes occur in the adenylation state of the accumulated RNA sequences, e.g., vitellogenic oocytes continue to accumulate poly(A)$^-$ RNAs but not poly(A)$^+$ RNAs but not poly(A)$^+$ RNAs; 2) qualitative (but not quantitative) changes in the poly(A)$^+$ RNA population take place; 3) degradation and resynthesis of poly(A)$^+$ RNA occur during vitellogenesis; 4) poly(A)$^+$ RNA is stable during vitellogenesis. These possibilities are not mutually exclusive and each is discussed below.

The possibility of changing states of RNA adenylation was examined in an analysis of the length of poly(A) tracts present at different oocyte stages (Cabada et al., 1977). The authors reported that two classes of poly(A)$^+$ RNA are present in oocytes; one class contains relatively long poly(A) tails ($\sim$ 60 A residues), and the other has relatively short poly(A) tails ($\sim$ 20 A residues). The two classes can be separated from each other by oligo(dT)-cellulose or poly(U)-Sepharose chromatography. Previtellogenic and early vitellogenic oocytes contain almost exclusively poly(A) with long tails while more mature oocytes contain both a larger proportion and amount of poly(A) with short tails. It was proposed that the former class of molecules is synthesized in young oocytes and is stable while the latter is continually accumulated during vitellogenesis. It is important to note, however, that we (unpublished results), and others (Dolecki and Smith, 1979), have not been able to detect the short poly(A) class at any stage of oogenesis.

The second possibility is that qualitative changes in the mRNA population occur during vitellogenesis. This has been investigated by examining the behavior of individual mRNA sequences throughout this period. cDNA clones, synthesized using poly(A)$^+$ RNA from oocytes as a template (Golden et al., 1980), were used as probes in Northern gel analysis of oocyte RNA. All RNA sequences of nuclear origin behaved similarly in that they reached a maximum level in previtellogenic or early vitellogenic oocytes and remained at this same level during vitellogenesis. The only exceptions were mitochondrial encoded poly(A)$^+$ RNAs which accumulate throughout oogenesis in a manner similar to 18S and 28S ribosomal RNA. For the majority of sequences examined (which are not very abundant) only oocyte poly(A)$^+$ RNA could be analyzed. In some cases, however, it was possible to detect a complementary sequence using total RNA. For these sequences as well, no substantial increase occurred during vitellogenesis, suggesting that poly(A)$^-$ [or short poly(A)] versions of these poly(A)$^+$ sequences are not accumulating during vitellogenesis. Similar conclusions have been drawn by other investigators studying identified sequences including histone mRNA (Van Dongen et al., 1981), and ribosomal protein mRNA (Pierandrei-Amaldi et al., 1982). In our study (Golden et al., 1980), only a few sequences were sufficiently abundant to permit an analysis of this kind, so this general conclusion for total RNA must be viewed as somewhat tentative. The conclusion that the poly(A)$^+$ RNA population undergoes few qualitative or quantitative changes

during vitellogenesis, however, is based upon measurements of many
more different sequences and is thus of more general applicability.
Also, in vitro translations of poly(A)$^+$ RNA isolated from different
stages of oogenesis do not show striking qualitative or quantitative
differences (Darnbrough and Ford, 1976).

The third issue concerns the role of RNA synthesis and
degradation during vitellogenesis. Work reported by Dolecki and
Smith (1979) suggests that in stage 3 and stage 6 oocytes, poly(A)$^+$
RNA is continually synthesized and degraded, albeit at very different
rates. These authors also demonstrate that the poly(A)$^+$ RNA present
in the early vitellogenic oocytes (stage 2) is very stable. These
two results--and the fact that stage 3 and stage 6 oocytes have the
same amount of poly(A)$^+$ RNA--are difficult to reconcile and make it
unclear whether transcripts synthesized early in oogenesis are stable
throughout oogenesis. These investigators have also shown that
poly(A)$^+$ RNA synthesized during the lampbrush chromosome stage is
transported to the cytoplasm and contains a significant fraction of
repeated sequences, similar to the RNA found in the cytoplasm of the
mature oocyte (Anderson et al., 1982). Possibly this is the fraction
of the oocyte poly(A)$^+$ RNA which is continually degraded and
resynthesized during oogenesis.

The fourth explanation is the stability (on a long time scale)
of oocyte poly(A)$^+$ RNA. These experiments, requiring in vivo
labeling of oocyte RNA, are difficult to perform due to the long
period of oocyte development and large but changing precursor pools.
In one study (Ford et al., 1977), it was shown that there are
poly(A)$^+$ RNA molecules which are synthesized very early in oogenesis
and are stable over a very long period of time with a half life of
greater than two years. It is not known, however, whether the
radioactive poly(A)$^+$ RNA was present in mature oocytes, or if it was
found only in arrested previtellogenic oocytes which did not proceed
through development. Dolecki and Smith (1979) proposed that the
immature oocyte does in fact remain at a holding point and does not
continue through vitellogenesis unless stimulated appropriately.
Since this in vivo labeling experiment (Ford et al., 1977) analyzed
RNA from entire ovaries rather than from fractionated cells, the
results obtained are compatible with this hypothesis.

Considering all of the evidence described above, the
interpretation we favor at present is that there are two classes of
poly(A)$^+$ RNA present in the stage 6 oocyte; one which is synthesized
early, is metabolically stable, and is stored during oogenesis for
translation after fertilization and another which is utilized for
translation by the oocyte during oogenesis and is subject to
metabolic turnover. In situ hybridization studies, which examine
oocytes at all stages of development, can be interpreted to support
this hypothesis (Capco and Jeffrey, 1982). There are two primary
sites of poly(A) localization in the mature oocyte, as assayed by

hybridization of poly(A)$^+$ RNA with ^{3}H-poly(U). These two sites appear at different times during oogenesis. Beginning at approximately stage 3, localization of grains is detectable as a ring around the germinal vesicle. As oogenesis continues, the ring moves to the periphery of the cell at the vegetal hemisphere. In stage 5 oocytes a second site of poly(A)$^+$ RNA localization is detectable, appearing in the interior cytoplasm of the cell, immediately under the vegetal side of the germinal vesicle. In the previtellogenic oocyte which synthesizes the metabolically stable poly(A)$^+$ RNA, no localization of poly(A)$^+$ RNA is detected. When the oocyte begins yolk uptake at stage 3, the visible poly(A)$^+$ RNA is segregated away from the site of synthesis (the germinal vesicle). As more yolk is deposited into the cell, the previtellogenic oocyte poly(A)$^+$ RNA is displaced further away from the nucleus to the cortical region of the cell. At least some newly synthesized poly(A)$^+$ RNA is not added to this pool. The new poly(A)$^+$ RNA is transported to the cytoplasm but remains separated from the poly(A)$^+$ RNA synthesized in previtellogenic oocytes by the yolk which is accumulated during vitellogenesis. Thus, the temporal appearance of these two sites is in rough agreement with the notion that there are two classes of poly(A)$^+$ RNA with different metabolic characteristics. The transition from previtellogenesis to vitellogenesis (the stage 2 to 3 transition), is important because it is the first time that poly(A)$^+$ RNA is detectable.

Studies addressing the patterns of protein synthesis during oogenesis also point to the stage 2/3 transition (the onset of vitellogenesis) as an important period in oocyte development. In vivo labeling studies reveal significant differences in protein synthesis as a function of oocyte stage (Harsa-King et al., 1979; Dixon and Ford, 1982). In one study (Harsa-King et al., 1979), 90 out of approximately 230 proteins (which can be resolved by two-dimensional electrophoresis) exhibit some change during oogenesis. In most cases these are quantitative changes, but in a few cases qualitative changes (the appearance or disappearance of new species) take place. Most of these qualitative changes occur between stages 2 and 3.

Maturation of the stage 6 oocyte requires hormonal stimulation and must occur for fertilization to take place. These cells, referred to as unfertilized eggs (UFEs), are radically different from oocytes in their biochemical and morphological characteristics. Among the most visible events which take place during maturation are the dispersal of the nuclear membrane (referred to as germinal vesicle breakdown), and the completion of meiosis up to the second meiotic metaphase. Because of the dramatic morphological changes associated with this period, it is of considerable interest to investigate the changes in gene activity which accompany oocyte maturation. Changes in the pattern of protein synthesis have been studied by several groups by in vivo labelling and two-dimensional gel electrophoresis (Ballantine et al., 1979; Bravo and Knowland, 1979; Meuler and Malacinski, this volume). The results of Ballantine et al. (1979) suggest that very few new (abundant) proteins are synthesized after maturation upon comparison with oocytes. Moreover, many of the most abundant proteins synthesized in oocytes are no longer visible among the labeled proteins from unfertilized eggs. The most obvious example of a protein in this latter category is actin. It is likely that ribosomal proteins also fall into this category, as it has been reported that ribosomal protein synthesis is dramatically reduced in the very early embryo by comparison with the oocyte (Pierandrei-Amaldi et al., 1982). The total amount of poly(A)$^+$ RNA also undergoes a change at oocyte maturation; it decreases by 40% from 1.8 ng in the stage 6 oocyte to 1.0 ng in the egg (Sugata et al., 1980).

RESULTS AND DISCUSSION

We have analyzed the behavior of individual mRNA sequences during maturation using Northern blots. Total RNA was prepared from mature oocytes and UFEs and hybridized to radioactive probes containing actin DNA, ribosomal protein DNA, or other random cDNA clones. The relative amount of each individual RNA did not change significantly at oocyte maturation (data not shown). This fits nicely with the observation that the amount of translatable actin mRNA, as assayed by comparing the in vitro translation products of oocytes and eggs (Sturgess et al., 1980), does not change significantly at maturation. The result also suggests that the decrease in the amount of poly(A)$^+$ RNA in the UFE as compared to the oocyte is not due to nonselective degradation of RNA sequences (i.e., 40% of each mRNA is degraded). Rather, it seems likely that some poly(A)$^+$ RNA molecules are deadenylated at oocyte maturation. To test this hypothesis, we compared the length of the poly tracts of individual mRNA sequences in mature oocytes to that in UFEs (Palatnik et al., 1979). Total RNA was bound to poly(U)-Sepharose and eluted with salt and increasing temperatures according to the method of Palatnik et al. (1979). In this way, it is possible to separate RNA molecules on the basis of their poly(A) tail length. The RNA in each fraction was analyzed by Northern gel analysis. Hybridization with a Xenopus actin cDNA probe (Schafer et al., 1982) shows that actin mRNA sequences are largely deadenylated during oocyte maturation (Fig. 1A). The same filters were rehybridized with a ribosomal protein cDNA clone; the ribosomal protein mRNA behaved like actin mRNA (Fig. 1C). The time of the deadenylation events correlates well with that of the changes in protein synthesis during this period of development, i.e., those specific mRNAs which are deadenylated are not translated in UFEs (Sturgess et al., 1980). A similar set of specific deadenylation events which are correlated with the lack of translation in vivo of specific mRNAs occurs at fertilization in the

OOCYTE   UFE
1 2 3 4 5 6 7 8 9 10 11   1 2 34 5 6 7 8 9 10 11

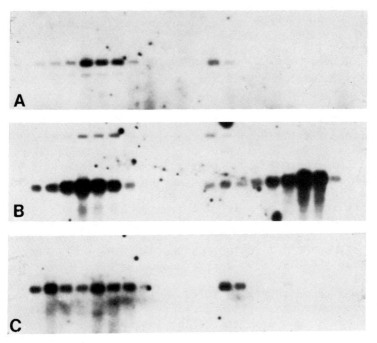

Figure 1: Thermal elution from poly(U)-Sepharose. Total RNA was prepared from mature oocytes and UFE (Colot and Rosbash, 1982). 1.8 mg of each RNA was hybridized to poly(U)-Sepharose in 0.7 M salt (Rosbash and Ford, 1974). RNA was then eluted with 0.1 M salt at 25°C and further eluted in a step-wise fashion by increasing the temperature in 2.5°C increments to a final temperature of 55°C. The remaining RNA was eluted in 90% formamide at 55°C. Each fraction was precipitated with ethanol and analyzed on a Northern blot. Hybridization was done as in the paper by Colot and Rosbash (1982). The blot was first hybridized to a nick-translated actin clone, XTC3-12 (Schafer et al., 1982). Audioradiography exposure time was 29 hr (1A). The filter was then boiled in dH_2O for 2 min and hybridized to XOC 2-7 (Colot and Rosbash, 1982) (1B); exposure time was 68 hr. The faint bands seen in the upper part of the gel are the result of a small amount of radiolabeled actin probe remaining on the filter. The same filter was boiled as above and rehybridized with radioactive Xom 102, a ribosomal protein cDNA clone for ribosomal protein Ll (Golden et al., 1980). This was exposed for approximately 84 hr. Lane 1 corresponds to RNA eluted at 25°C; lane 2 = 35°C eluate; lane 3 = 37.5°C eluate; lane 4 = 40°C eluate; lane 5 = 42.5°C eluate; lane 6 = 45.0°C eluate; lane 7 = 47.5°C eluate; lane 8 = 50.0°C eluate; lane 9 = 52.5°C eluate; lane 10 = 55.0°C eluate; lane 11 = 55°C eluate in 90% formamide.

surf clam, Spisula (Rosenthal et al., 1983; and Ruderman et al., this volume).

It must be the case that many oocyte sequences do not behave in this manner since the egg contains approximately 50% as much poly(A)$^+$ RNA as the oocyte (Sagata et al., 1980). Also, in vivo protein synthesis data indicate that there are many mRNA sequences present on polysomes in UFEs; indeed, the fraction of ribosomes in polysomes increases at maturation (Woodland, 1974), and the rate of protein synthesis goes up as well (Smith, 1975). Since there is a correlation between mRNA deadenylation and the lack of template activity in vivo, sequences which are actively translated at (or shortly after) fertilization are probably poly(A)$^+$. Moreover, Sagata and coworkers (1980) found that poly(A)$^+$ RNA isolated from UFEs has a longer poly(A) tract than the poly(A)$^+$ RNA from mature oocytes. Our data are consistent with this observation in that the poly(A)$^+$ RNA population from UFEs is eluted from poly(U)-Sepharose at higher temperatures than the poly(A)$^+$ RNA from mature oocytes (Fig. 2). All of these results taken together suggest that some mRNA sequences, like actin and ribosome proteins, are deadenylated while other sequences, like XOC 2-7 (Colot and Rosbash, 1982; Fig. 1B), are adenylated at the same time during oocyte maturation.

The data shown in Fig. 1 and 2 suggest that the adenylation and deadenylation which take place during oocyte maturation occur in a sequence-specific manner. It is also possible that any given RNA sequence is either predominantly adenylated or deadenylated at this time. In the case of actin and ribosomal protein mRNA, the deadenylation events are closely correlated with a cessation of translation. Whether the deadenylation is responsible for the translational arrest, or is a consequence of the translational arrest, or is not directly related to the rate of protein synthesis is not known.

The series of events which immediately follow fertilization occurs in the absence of de novo transcription, which does not begin until the 12th cleavage division (Newport and Kirschner, 1982), [referred to as the midblastula transition (MBT)]. Therefore, maternal mRNAs must be translated until this time. For this reason it is interesting to follow the fate of maternal mRNA sequences during early embryogenesis, especially prior to the MBT. Sagata and coworkers (1980) investigated the changes in the poly(A)$^+$ RNA population during early embryogenesis. These experiments demonstrated that between fertilization and blastula there is a steady increase in the per embryo level of poly(A)$^+$ RNA; at blastula it reaches the level present in the stage 6 oocyte, 1.8 ng/embryo. As mentioned above, no new transcription occurs during this time, so adenylation of maternal RNA must occur prior to the MBT.

We have studied the behavior of the XOC 2-7 RNA sequence after

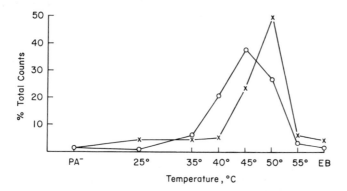

Figure 2:   Thermal elution of oocyte and UFE RNA and hybridization
with ³H-poly(U). 50 μg of RNA was bound to poly(U)-Sepharose.
Fractions were eluted as described in Fig. 1. Each fraction was
hybridized to ³H-poly(U) as described in Rosbash and Ford (1974).
Poly(A)⁻ RNA is RNA which did not bind to poly(U)-Sepharose. EB is
the final eluate, i.e., RNA which was eluted in 90% formamide at
55°C.

fertilization. The poly(A)⁺ RNA is not further adenylated; instead,
it is progressively deadenylated. As cleavage continues, the
sequence disappears, as we are barely able to detect it by Northern
gel analysis (Fig. 3A). Although XOC 2-7 RNA is adenylated at
maturation, it and similarly expressed sequences cannot be
contributing to the increase in poly(A)⁺ RNA during cleavage. It is
possible that the increase is due to the readenylation of sequences
deadenylated at oocyte maturation (Sagata et al., 1980). However,
actin and ribosomal protein RNAs are deadenylated at oocyte
maturation but not readenylated during cleavage. On the contrary,
these sequences decrease dramatically in abundance as is shown for a
ribosomal protein mRNA in Fig. 3B. These results agree with the in
vivo protein synthesis data of Sturgess et al. (1980), which show
that there is little or no actin protein synthesis in either the
preblastula embryo or the unfertilized egg. Since actin mRNA levels,
measured by in vitro translation, do not change at fertilization, it
has been proposed that the regulation of actin synthesis during very
early embryogenesis takes place at the translational level. The data
presented here suggest that the regulation is indeed translational in
unfertilized eggs, but that the actin mRNAs are largely degraded soon
after fertilization. Thus there is sequence specific mRNA
degradation during the earliest stages of embryogenesis.
Deadenylation of RNA precedes (in the case of actin and ribosomal
proteins mRNA), or accompanies (in the case of XOC 2-7 RNA), the
degradation of the RNA.

1 2 3 4 5 6 7 8 9 10 11 12 13   1 2 3 4 5 6 7 8 9 10 11 12 13

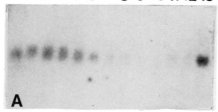
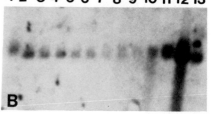

A   B

Figure 3: Hybridization to early embryo RNA. 12 μg of total RNA was
prepared from each embryo stage. Northern gel analysis was done as
described by Colot and Rosbash (1982).

3A) ^{32}P-labeled XOC 2-7 DNA was used as a probe and the blot was
exposed for 16 hrs.

3B) ^{32}P-labeled M13-Xom 92 DNA, a ribosomal protein cDNA clone
(corresponding to ribosomal protein L14), subcloned into the
single-stranded phage vector M13 mp8 was hybridized to the same
nitrocellulose filter as in 3A.

Lane 1, mature oocyte RNA.   Lane 8, stage 10-11 RNA.
Lane 2, UFE RNA.   Lane 9, stage 11-12 RNA.
Lane 3, stage 1-2 RNA.   Lane 10, stage 13-14 RNA.
Lane 4, stage 3-4 RNA.   Lane 11, stage 15-16 RNA.
Lane 5, stage 5-6 RNA.   Lane 12, tadpole RNA, stage 42.
Lane 6, stage 7-8 RNA.   Lane 13, total ovary RNA.
Lane 7, stage 9-10 RNA.

     Since none of the three RNAs discussed above are adenylated
during cleavage, the question remains as to which RNAs contribute to
this increase in the poly(A)$^+$ RNA. Perhaps there are poly(A)$^-$
sequences present in the mature oocyte which are adenylated during
cleavage and account for the observed increase in poly(A)$^+$ RNA.
These RNAs might accumulate during vitellogenesis and constitute the
second class of mRNA made during oogenesis.

     In this report we have reviewed the control of gene expression
during oogenesis and early embryogenesis in terms of mRNA
accumulation, adenylation, turnover, and translation. As the embryo
continues to develop, additional mechanisms which control gene
expression may be utilized. For example, the first zygotic
transcription products detected include tRNA, 5S RNA, and snRNAs. In
the case of tRNA and 5S RNA, translation is directly influenced and,
in the case of snRNAs, RNA processing may be affected. All of these
RNAs directly influence translation in the case of the tRNA and 5S

RNA, and possibly RNA processing in the case of the snRNAs (Lerner and Steitz, 1981; Zieve, 1981). It has been reported that there is a pool of snRNA binding proteins in Xenopus oocytes (DeRobertis et al., 1982). DeRobertis and coworkers injected snRNA from HeLa cells into Xenopus oocytes and found that the RNA was incorporated into ribonucleoprotein particles (snRNPs). They also found that the assembled snRNAs are transported to and sequestered in the nucleus. The RNA binding proteins are cytoplasmic and, although they have not been titrated, are present in sufficient quantity to bind at least the equivalent of snRNA from 1,000 HeLa cells. The presence of this large amount of snRNA binding protein suggests that there might be a deficiency of snRNA in mature oocytes. In this light, it is not surprising that the prominent synthesis of snRNA is among the first zygotic transcription products. (Newport and Kirschner, 1982). Because the snRNA complex might be required for mRNA processing, it is possible that the mature oocyte is relatively inefficient at splicing precursor RNAs into mature mRNAs relative to embryos.

Although oogenesis and early embryogenesis in Xenopus laevis are complex processes involving a wide array of controls, it is possible to examine key events during development in order to learn which mechanisms are involved. The accumulation of RNA molecules takes place in the previtellogenic and early vitellogenic oocyte. The maturation of the oocyte triggers some biochemical events which directly affect translation and mRNA degradation. Not until the 12th cleavage division does the embryo begin de novo transcription. The midblastula transition is, therefore, the important switch from maternal to zygotic gene expression. Most experiments performed to date have addressed transcriptional and translational aspects of gene expression. Recent advances have defined and examined mRNA processing. It should be possible and interesting to compare the ability of oocytes and embryos to use mRNA processing as a means of controlling gene expression.

ACKNOWLEDGEMENTS

We are grateful to our colleagues Mark Gray, Alain Vincent, and Michael Wormington for comments on this manuscript. This work was supported by a grant from the NIH (HD 08887) to MR.

REFERENCES

Anderson, D.M., Richter, J.D., Chamberlin, M.E., Price, D.H., Britten, R.J., Smith, L.D., and Davidson, E.H. (1982). Sequence organization of the poly(A) RNA synthesized and accumulated in lampbrush chromosome stage Xenopus laevis oocytes. J. Mol. Biol. 155:281-309.

Ballantine, J.E.M., Woodland, H.R., and Sturgess, E.A. (1979). Changes in protein synthesis during development of Xenopus laevis. J. Embryol. Exp. Morphol. 51:135-153.

Bravo, R. and Knowland, J. (1979). Classes of proteins synthesized in oocytes, eggs, embryos and differentiated tissues of Xenopus laevis. Differentiation 13:101-108.

Cabada, M.O., Darnbrough, C., Ford, P.J., and Turner, P.C. (1977). Differential accumulation of two size classes of poly(A) associated with messenger RNA during oogenesis in Xenopus laevis. Develop. Biol. 57:427-439.

Capco, D.G. and Jeffrey, W.R. (1982). Transient localizations of messenger RNA in Xenopus laevis oocytes. Develop. Biol. 89:1-12.

Colot, H.V. and Rosbash, M. (1982). Behavior of individual maternal poly(A)+ RNAs during embryogenesis of Xenopus laevis. Develop. Biol. 94:79-86.

Darnbrough, C. and Ford, P.J. (1976). Cell-free translation of messenger RNA from oocytes of Xenopus laevis. Develop. Biol. 50:285-301.

Davidson, E.H. (1976). "Gene Activity in Early Development." New York/San Francisco/London, Academic Press.

DeRobertis, E.M., Lienhard, S., and Pavisot, R.E. (1982). Intracellular transport of microinjected 5S and small nuclear RNAs. Nature 295:572-577.

Dixon, L.K. and Ford, P.J. (1982). Regulation of protein synthesis and accumulation during oogenesis in Xenopus laevis. Develop. Biol. 93:478-497.

Dolecki, G.J. and Smith, L.D. (1979). Poly(A)+ RNA metabolism during oogenesis in Xenopus laevis. Develop. Biol. 69:217-236.

Dumont, J.N. (1972). Oogenesis in Xenopus laevis: I. Stages of oocyte development in laboratory maintained animals. J. Morphol. 136:153-180.

Ford, P.J., Mathieson, T., and Rosbash, M. (1977). Very long-lived messenger RNA in ovaries of Xenopus laevis. Develop. Biol. 57:417-426.

Golden, L., Schaffer, U., and Rosbash, M. (1980). Accumulation of individual poly(A)+ RNAs during oogenesis of Xenopus laevis. Cell 22:835-844.

Harsa-King, M.L., Bender, A., and Lodish, H.F. (1979). Stage specific changes in protein synthesis during Xenopus oogenesis. In "Eukaryotic Gene Regulation," R. Axel, T. Maniatis, and C.F. Fox (eds.), Vol. XIV. ICN-UCLA Symposia on Molecular and Cellular Biology. Academic Press, New York.

Keem, K., Smith, L.D., Wallace, R.A., and Wolf, D. (1979). Growth rate of oocytes in laboratory-maintained Xenopus laevis. Gametes Res. 2:125-135.

Lerner, M.R. and Steitz, J.A. (1981). Snurps and Scyrps. Cell 25:298-300.

Newport, J. and Kirschner, M. (1982). A major developmental transition in early Xenopus embryos: II. Control of the onset of transcription. Cell 30:687-696.

Palatnik, C.M., Storti, R.V., and Jacobson, A. (1979). Fractionation and functional analysis of newly synthesized and decaying messenger RNAs from vegetative cells of Dictyostelium discoideum. J. Mol. Biol. 128:371-395.

Pierandrei-Amaldi, P., Campioni, N., Beccari, E., Bozzoni, I., and Amaldi, F. (1982). Expression of ribosomal-protein genes in Xenopus laevis development. Cell 30:163-171.

Rosbash, M. and Ford, P.J. (1974). Polyadenylic acid-containing RNA in Xenopus laevis oocytes. J. Mol. Biol. 85:87-101.

Rosenthal, E.T., Tansey, T.R., and Ruderman, J.V. (1983). Sequence-specific adenylations and deadenylations accompany changes in the translation of maternal mRNA after fertilization of Spisula oocytes. J. Mol. Biol., in press.

Sagata, N., Shiokawa, K., and Yamana, K. (1980). A study on the steady state population of poly(A)$^+$ RNA during early development of Xenopus laevis. Develop. Biol. 77:431-448.

Schafer, U., Golden, L., Hyman, L.E., Colot, H.V., and Rosbash, M. (1982). Some somatic sequences are absent or exceedingly rare in Xenopus oocyte RNA. Develop. Biol. 94:87-92.

Smith, L.D. (1975). Molecular events during oocyte maturation. In "Biochemistry of Animal Development," R. Weber (ed.), Vol. III. New York, Academic Press.

Sommerville, J. (1977). Gene activity in the lampbrush chromosomes of amphibian oocytes. In "Biochemistry of Cell Differentiation," J. Paul (ed.), II, pp. 79-156. Baltimore, University Park Press.

Sturgess, E.A., Ballantine, J.E.M., Woodland, H.R., Mohun, P.R., Lane, C.D., and Dimitriadis, G.J. (1980). Actin synthesis during the early development of Xenopus laevis. J. Embryol. Exp. Morphol. 58:303-320.

Van Dongen, W., Zaal, R., Moorman, A., and Destree, O. (1981). Quantiation of the accumulation of histone messenger RNA during oogenesis in Xenopus laevis. Develop. Biol. 86:303-314.

Woodland, H.R. (1974). Changes in the polysome content of developing Xenopus laevis embryos. Develop. Biol. 40:901-101.

Zieve, G.W. (1981). Two groups of small stable RNAs. Cell 25:296-297.

PROTEIN SYNTHESIS PATTERNS DURING

EARLY AMPHIBIAN EMBRYOGENESIS

Debbie Crise Meuler and George M. Malacinski[*]

SUMMARY

Changes in two-dimensional (2-D) gel protein synthesis patterns during the development of Ambystoma mexicanum and Xenopus laevis embryos were cataloged. Proteins which appear to be synthesized only during specific developmental stages were identified and a comparison of those results was made to similar types of analyses previously reported in the literature. It was concluded that neither species displayed changes in 2-D gel patterns during oocyte maturation. At fertilization, however, synthesis of some oogenic proteins ceased temporarily but resumed later in development. In Xenopus eggs the synthesis of several novel proteins was detected at fertilization. Later, during gastrulation, several additional novel proteins were also synthesized. During axolotl gastrulation, even more dramatic changes occurred. Ten percent of the proteins detected at earlier stages ceased synthesis at gastrulation. Furthermore, 10% of the proteins synthesized at gastrulation were novel. Some of them are specific to different tissues of the embryo. The actins and tubulins were also identified on the 2-D gels and their synthesis was monitored during axolotl development.

———————

[*]From the Program in Molecular, Cellular, and Developmental Biology, Department of Biology, Indiana University, Bloomington, IN 47405.

267

INTRODUCTION

Amphibian embryos have historically been employed as a model
system for analyzing such concepts as "determination" and "cell
differentiaton." Their relatively large size, external development,
and resilient response to surgical manipulation are among their key
features that have been exploited by several generations of
embryologists. In contrast to several other types of embryos being
discussed at this symposium, amphibian embryos are transcriptionally
dormant from fertilization up through the first 9-10 rounds of
mitosis (Gurdon and Woodland, 1968; Nakahashi and Yamana, 1974;
Newport and Kirschner, 1982; Frost and Malacinski, unpublished
observations). Yet during that period of post-fertilization
embryogenesis, several important events of early pattern
specification occur. The dorsal/ventral polarity of the egg is
established, the animal/vegetal cleavage pattern becomes fixed, the
blastocoel arises in the animal hemisphere, and the germplasm becomes
segregated into a few vegetal hemisphere cells. No doubt other as
yet unidentified pattern formation events also occur. Although those
pattern specification events apparently do not require substantial
amounts of gene transcription, protein synthesis is definitely
required. From the beginning of the present era of biochemical
embryology it has been known that protein synthesis is required for
early amphibian embryogenesis to proceed in a normal fashion (Brachet
et al., 1964).

It appears, therefore, that early amphibian pattern formation is
predominantly controlled at the level of translation rather than at
the level of transcription. Accordingly, the studies described in
the present report were undertaken to gain a better understanding of
the general features of the translational controls that may operate
during early embryogenesis. C^{14} amino acid labeling, tissue
dissection, and two-dimensional (2-D) gel electrophoresis was
employed to analyze the protein synthesis patterns of early
embryogenesis. Rather than confining the present studies to only the
period between fertilization and the mid-blastula transition (when
transcription is initiated; Newport and Kirschner, 1982), a
comprehensive survey of protein synthesis patterns from oogenesis
through the early stages of organogenesis was carried out.
Initially, a general survey of protein synthesis patterns was made.
The synthesis patterns in the dorsal or ventral regions of early
embryos then were established, and finally the synthesis of a few
well characterized proteins (e.g., actin and tubulin) was monitored.

The catalog that has been compiled will be employed in future
studies directed toward understanding the roles such events as egg
activation and tisssue interactions play in controlling the pattern
of protein synthesis.

MATERIALS AND METHODS

Ambystoma mexicanum embryos were obtained from natural
spawnings, manually dejellied, and staged according to Bordzilowskaya
and Dettlaff (1979). Xenopus laevis embryos were obtained from
artificial insemination of hormone stimulated females, chemically
dejellied, and staged according to Nieuwkoop and Faber (1956).
Proteins synthesized during a specific developmental stage were
labeled by microinjecting at the beginning of the desired stage a
mixture of C^{14}-labeled amino acids derived from a yeast protein
hydrolysate (Amersham #CFB.25) (Fig. 1). The embryos were then
incubated at room temperature for the 4-6 hr that usually
corresponded to the length of the developmental stage under
examination. Newly synthesized proteins were extracted, with the aid
of sonication, from whole embryos at the various stages up to
neurulation. From neurulation onward, proteins were extracted from
embryos dissected into dorsal and ventral halves. Sonication was
used in the extraction procedure because it allowed for maximum
recovery of labeled proteins from small quantities of tissue.

Proteins were separated by the two-dimensional gel
electrophoresis method described by O'Farrell (1975). The
first-dimension gel was an isoelectric focusing gel, using a
combination of pH 3-10 and pH 5-7 LKB ampholines in the ratio of 1:4.
The pH gradient of the first-dimension gel was determined by slicing
the gels into 0.5 cm slices and soaking the slices for 5-10 min in
2.0 ml of a 9.5 M urea solution prepared with degassed, distilled
water. A pH determination was then made with a standard pH meter.
The pH gradient, although not determined for each first-dimension
gel, was checked periodically and found to be approximately 5.0-7.2.

In the second dimension, a concentration of 10% acrylamide was
used. That concentration of acrylamide was found to provide optimal
separation of the proteins. At each developmental stage,
approximately 270 proteins were resolved. Because the buffer in
which the proteins were extracted contained substances that render
standard protein assays unreliable, it was difficult to standardize
the sample size according to the protein concentration. Therefore,
approximately 25,000-30,000 TCA precipitable cpm were routinely
loaded on each gel. That number of cpm represented the extraction
products of approximately 2-4 embryos.

The gels were prepared for fluorography by placing them in 1.0 M
sodium salicylate for 20 min and then drying them under vacuum
(Chamberlain, 1979). The dried gels were packed under Kodak X-Omat
X-ray film for approximately four weeks.

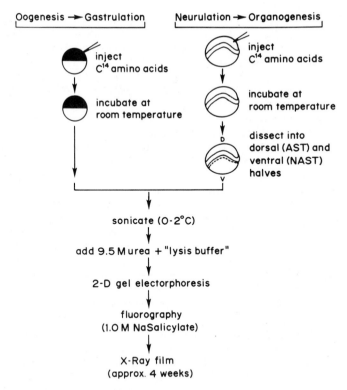

<u>Figure 1</u>: Method used to identify the newly synthesized proteins in
specific developmental stages of axolotl development. The same
method was used in analyzing <u>Xenopus</u> embryo protein patterns except
that embryos after gastrulation were not dissected into dorsal and
ventral halves. Approximately 0.1 μCi of a lyopholyzed C^{14} amino
acid mixture was microinjected into each embryo. The proteins were
homogenized on ice with a 30-sec burst from a Bransen J-17A
sonicator for 30 sec.

RESULTS

<u>Review of the Methodology and Criteria Used to Detect Changes in
Synthesis Patterns</u>

Initially, representative fluorographs of the newly synthesized
proteins of various developmental stages were compared. Each of the
changes in the pattern of protein synthesis were cataloged. In most
cases the electrophoretic patterns, even from identical developmental
stages, were not completely superimposable because of slight (and
uncontrollable) variations during the separation of the proteins.
For comparisons, the fluorographs were therefore oriented with the

aid of several of the major proteins synthesized at all developmental
stages.

Changes in the pattern of protein synthesis were considered
significant if they were 1) present on duplicate fluorographs of the
appropriate developmental stage and 2) recognizable on fluorographs
which contained mixed samples. In the first instance, several
fluorographs prepared from embryos of different spawnings were
analyzed for reproducibility. Some differences (e.g., 4-5 spots)
were found in the 2-D gel patterns from different spawnings. These
proteins displayed variations from spawning to spawning and were
deleted from the analysis described below. Conversely, any change in
the pattern of protein synthesis was considered significant only if
it appeared in several fluorographs prepared from the same stage of
different spawnings. This approach discounted the possibility that
changes in the pattern of protein synthesis might be an artifact of
different spawnings.

In the second instance, it was important to determine whether
changes in the pattern of protein synthesis might be due to artifacts
of the extraction procedure. Embryos labeled with C^{14} amino acids at
different developmental stages were therefore co-extracted and
analyzed on single fluorographs. The resulting gel patterns always
represented a composite of the changes previously noted when both
stages were individually compared. This observation indicated that
the changes were not due to a recovery artifact resulting from a
stage specific difference in extraction efficiencies.

For brevity, the terms synthesis "began" or "ceased" were used
in reference to the appearance or disappearance of a protein from a
fluorograph, not necessarily the actual onset or termination of
synthesis of a protein. "Overloading" a 2-D gel with cpm increases
the resolving capabilities of the gel which might indicate whether
the synthesis of a protein had actually ceased or had only fallen
below the limit of resolution of the gel system. Overloading a
fluorograph, however, resulted in a large amount of streaking which
made analyzing the fluorographs difficult. Therefore, gel
overloading to detect exceedingly low levels of synthesis was not
included in the present studies.

Changes in the Pattern of Protein Synthesis During Oocyte Maturation

Mexican axolotl (Ambystoma mexicanum) and Xenopus laevis oocytes
were manually removed from the ovary and sorted according to size.
Axolotl oocytes were staged according to a Xenopus oocyte staging
series (Dumont, 1972) since an axolotl oocyte staging series is not
available. Stage VI oocytes, with distinct animal and vegetal poles,
were collected from axolotl and Xenopus females. Half of the Stage
VI oocytes were induced to mature in vitro by the addition of 0.3 µM
progesterone. Maturation was scored by the appearance of a white

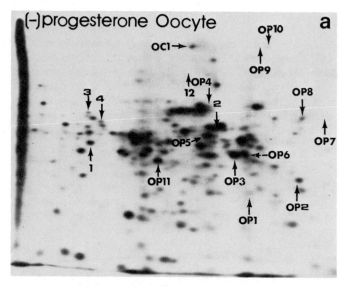

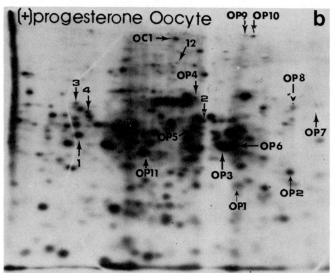

Figure 2: Fluorographs of 2-D gel separations of proteins
   synthesized during different stages of axolotl development. The
   proteins were first separated by isoelectric focusing on a pH
   gradient of 5.0 (right) to 7.2 (left), followed by separation using
   a denaturing 10% acrylamide gel. a) Stage VI oocytes injected and
   incubated for 5 hr. b) Stage VI oocytes induced to mature with 0.3
   μM progesterone, injected and incubated for 5 hr until maturation.

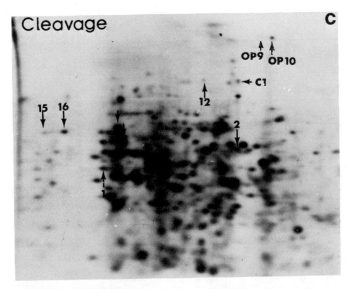

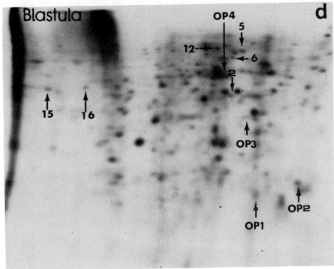

Figure 2 (continued):

c) Two-cell stage embryos injected and incubated for 6 hr to 32-cell stage. d) Blastula stage embryos injected at stage 8 and incubated for 5 hr to stage 9. Arrows indicate major difference between these stages. Protein #12 was observed on the original fluorograph of a but was difficult to detect on photographs of the fluorograph.

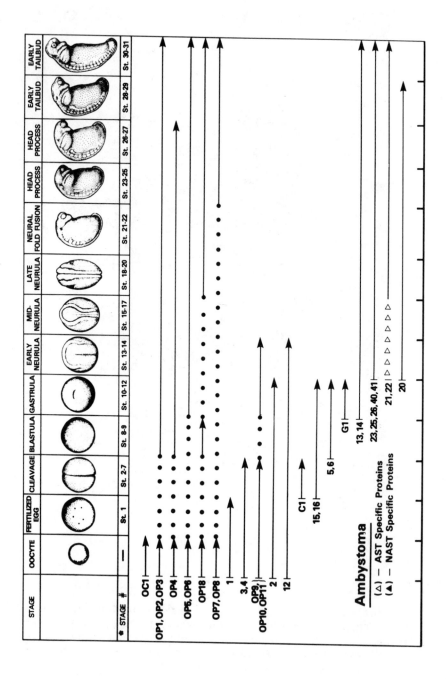

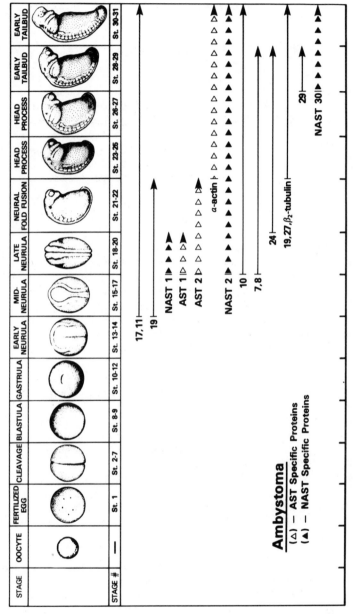

Figure 3: Summary of protein synthesis patterns which change during early axolotl development. a) Oogenesis to early neurula (St. 13/14). b) Early neurula (St. 15/17) to early tailbud (St. 30/31). Solid lines indicate periods of synthesis of the protein, and dotted lines indicate periods when the synthesis of the protein cannot be detected. Embryos staged according to Bordzilowskaya and Dettlaff (1979).

spot at the animal pole which indicated the release of the first
polar body. Untreated Stage VI and hormone maturing oocytes were
injected with C^{14}-labeled amino acids and incubated for 5-6 hr at
room temperature. The proteins were extracted and analyzed as
described in the Materials and Methods.

The proteins synthesized in axolotl oocytes before and during
hormone induced maturation are shown in Fig. 2. Neither axolotl nor
Xenopus oocytes displayed changes in protein synthesis during oocyte
maturation (only fluorographs of axolotl oocyte proteins are shown).
The process of oocyte maturation, including germinal vesicle
breakdown, apparently does not require the synthesis of new proteins
as detected by gel electrophoresis. These results agree with work
previously published by Ballantine et al. (1979). They found no
consistent change in the pattern of protein synthesis of
S^{35}-methionine labeled proteins in Xenopus oocytes before and after
maturation.

Protein Synthesis After Fertilization

In both Xenopus and axolotl eggs, several proteins synthesized
during oogenesis were not detected as labeled spots after
fertilization. Proteins of this type are marked on Fig. 2 and 3 (for
the axolotl) and Fig. 4 (for Xenopus). The synthesis of those
proteins that were first labeled during oogenesis, however,
eventually resumed. In axolotl embryos the synthesis of OP1, OP2,
OP3, and OP4 resumed at blastulation. The synthesis of OP5 and OP6
was reinitiated during gastrulation while OP7 and OP8 were
synthesized again at neurulation (St. 23/25). In Xenopus embryos,
synthesis of OP3, OP15, OP16, and OP17 resumed at the onset of
cleavage, while the synthesis of OP12 began during blastulation.

These results agree, for the most part, with results previously
reported by Ballantine et al. (1979) and Bravo and Knowland (1979).
Those workers also detected the cessation of synthesis of a subset of
Xenopus oogenic proteins after fertilization and subsequent
resumption of synthesis at later stages. Ballantine et al. (1979)
reported that most of those proteins reappeared at gastrulation.
Bravo and Knowland (1979), however, found most of those proteins
reappeared at the onset of cleavage, while one of them reappeared at
the blastula stage.

In the present studies, it was also discovered that during
fertilization of Xenopus eggs several proteins were synthesized for
the first time. Those proteins represent novel non-oogenic proteins
and are coded in Fig. 4 for Xenopus. The synthesis of those novel
non-oogenic proteins at fertilization in Xenopus continued during all
the later stages of development. The synthesis of such novel
non-oogenic proteins at fertilization in Xenopus was not detected in
earlier studies of either Ballantine et al. (1979), Bravo and

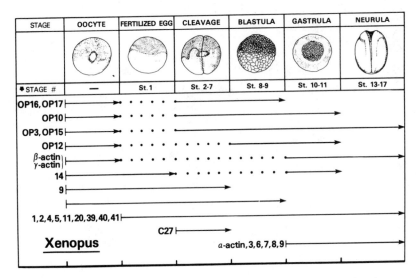

Figure 4: Summary of protein synthesis patterns which change during Xenopus development from oogenesis through early neurula. Solid and dotted lines key same as in Fig. 4. Xenopus embryos staged according to Nieuwkoop and Faber (1956).

Knowland (1979), or in the present study of axolotl protein synthesis patterns.

Newly Synthesized Proteins at the Onset of Cleavage

In axolotl eggs the synthesis of several proteins, not detectable during oogenesis, began at the onset of cleavage. For example, C1, #15, and #16 are indicated in Fig. 2c. As can be seen in Fig. 2 and 3, the synthesis of C1 was detected only during cleavage, while #15 and #16 were synthesized during both cleavage and blastula. Those proteins are synthesized during the developmental period when the embryo undergoes rapid cell divisions.

As indicated in Fig. 4, the synthesis of one Xenopus protein (C27) began during cleavage and was specific for that stage. Ballantine et al. (1979) reported no change in the 2-D gel patterns of fertilized versus cleaving embryos. Bravo and Knowland (1979), however, did report that several novel non-oogenic proteins appeared at the onset of cleavage.

Proteins Synthesized at Gastrulation

An unusual change in the 2-D gel pattern was observed at gastrulation. Careful examination of Fig. 5 revealed that approximately 10% of the proteins synthesized in previous stages were

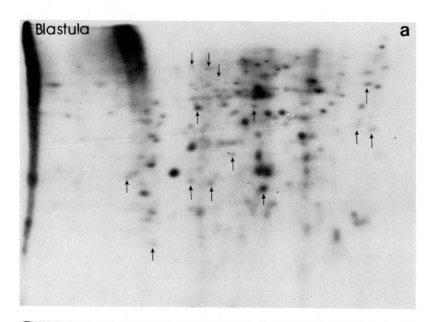

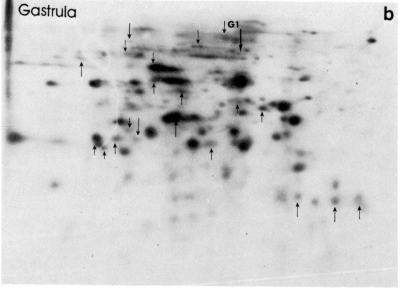

Figure 5: Fluorographs of 2-D gel separations of proteins
 synthesized during axolotl blastulation and gastrulation.
 a) Blastula stage embryos injected at stage 8 and incubated for 5
hr to stage 9. b) Gastrula stage embryos injected at stage 10 and
incubated for 6 hr to stage 11. Arrows in (a) indicate proteins
that were found in previous stages but disappear at gastrulation.
Arrows in (b) reveal the synthesis of proteins that begin at
gastrulation.

not found at the onset of gastrulation. Those proteins apparently
were either not synthesized at all during gastrulation or were
synthesized at levels below the limits of resolution of the system.
It was also discovered that 10% of the proteins detected at
gastrulation are novel non-oogenic proteins. The synthesis of those
proteins was not detected at earlier stages. Their synthesis
continued during all later stages of development. At a time when the
axolotl embryo undergoes drastic changes in morphology, it apparently
also undergoes substantial changes in protein synthesis.

In Xenopus, a similar situation was not observed. Several
proteins appeared at gastrulation (Fig. 4: #3, #6, #7, #8, and #9)
that represent new, non-oogenic proteins. Those proteins, however,
represent only a minor percentage of the total number of proteins
detected during Xenopus gastrulation. Furthermore, the loss of 10%
of those proteins, detected in previous stages, does not occur.
Ballantine et al. (1979) reported the synthesis of a few novel
proteins at gastrulation. Yet Bravo and Knowland (1979) did not find
any new proteins synthesized at gastrulation. The dramatic change in
the pattern of protein synthesis at gastrulation, therefore, is found
only in axolotl and not in Xenopus embryos.

Proteins Synthesized in Different Tissue Layers

Changes in the pattern of protein synthesis in various tissues
of post-gastrula stage axolotl embryos were analyzed. After embryos
were injected with C^{14} amino acids they were incubated and then
dissected into dorsal and ventral halves (Fig. 1): The dorsal half
of the embryo--consisting of presumptive axial structures including
the neural tube, notochord, and somites--is designated as AST (axial
structure tissue). The ventral half of the embryo--consisting of
yolky endoderm that develops into gut and related tissues--is
designated as NAST (non-axial structure tissue).

As summarized in Fig. 3, for most axolotl stages examined, novel
non-oogenic proteins were detected. Most of these proteins (i.e.,
#13, #14, #19, or #23) were found in all of the later stages. Some
proteins (for example, #20) were detected from St. 13/14 (early
neurula) through St. 28/29 (early tailbud), but not at St. 30/31.

Several of those newly synthesized proteins were found to be
specific to one or the other half of the embryo. The first dorsal
specific proteins were detected during early neurula (St. 13/14).
Proteins coded #21 and #22 in Fig. 3 were found in the AST during
early to mid-neurulation. Later they were found in all regions of
the embryo. Two additional AST specific proteins were detected
during neural fold fusion. AST #1 was detected during St. 18/20 and
AST #2 was detected during St. 18/20 and St. 21/22. At St. 23/25
(head process stage), yet another AST specific protein was found.
That protein was identified as α-actin (see below).

The synthesis of two NAST specific proteins was detected during late neurula (St. 18/20). NAST #1 was specific to St. 18/20, while NAST #2 was detected at all later stages. It was not until early tailbud (St. 28/29) that another NAST specific protein was recognized (NAST #30).

Ballantine et al. (1979) found several proteins specific to different regions of Xenopus embryos. The first "ectoderm" specific protein was detected as early as blastulation. Several more appeared during neurulation and early tailbud. The first "endoderm" specific protein was not detected until early neurulation. More appeared around early tailbud. Bravo and Knowland (1979), however, did not include an analysis of proteins synthesized in specific regions of early embryos.

Identification of Specific Proteins

Relatively few proteins seen on 2-D gels can be equivocally identified without detailed characterization procedures. However, two proteins--actin and tubulin--are well characterized and readily identifiable on 2-D gels. Three classes of actins are presently known: γ- and β-actins are found in the cytoplasm of all cells, α-actin is specific to muscle cells. On the axolotl gels the position of actin was recognized by comparing peptide maps from chymotrypsin digests (Cleveland et al., 1977) of a rabbit muscle actin standard to the presumed actin spot cut out of a fluorograph.

As mentioned in Fig. 3 and 6, γ- and β-actins were synthesized in all stages of development and probably in all tissues as well. Conversely, the synthesis of α-actin was first detected during St. 23/25 (head process stage) and was restricted to the AST. Mohun et al. (1980) have also reported that the synthesis of α-actin begins at the head process stage of development, in the ectodermal half of the embryo.

In Xenopus, actin synthesis apparently does not follow that pattern. Both Ballantine et al. (1979) and Sturgess et al. (1980) examined actin synthesis in Xenopus embryos and found a different pattern. The synthesis of γ- and β-actin was detected during oogenesis, but not immediately after fertilization. Their synthesis resumed as blastulation. α-actin synthesis, on the other hand, was detected as early as gastrulation and was specific to the "ectodermal" region. This difference in actin synthesis between the two species has several implications that will be discussed later.

As mentioned previously, tubulins are another class of proteins that are well characterized and are easily identifiable on 2-D gels. In amphibians there are four types of tubulin subunits: α_1, α_2, β_1, and β_2. The position on fluorographs of the four tubulin subunits was based on their isoelectric points and molecular weights. Those

Stage / Protein	Oogenesis	Fertilized Egg	Cleavage	Blastula	Gastrula	Early Neurula	Neural Fold Fusion	Head Process	Tailbud
α — actin	—	—	—	—	—	—	—	+	+
β — actin	+	+	+	+	+	+	+	+	+
γ — actin	+	+	+	+	+	+	+	+	+
α_1 — tubulin	+	+	+	+	+	+	+	+	+
α_2 — tubulin	+	+	+	+	+	+	+	+	+
β_1 — tubulin	+	+	+	+	+	+	+	+	+
β_2 — tubulin	—	—	—	—	—	—	—	+	+

Figure 6: Summary of the changes in actin and tubulin synthesis
during axolotl early development: (+) indicates detectable
synthesis and (−) indicates absence of detectable synthesis.

tubulin identities were confirmed by comparing our fluorographs with
those published by Mohun et al. (1980) which contained detailed
identifications of the tubulin subunits. In axolotl embryos, α_1, α_2,
and β_1 were found during all stages of development. β_2 synthesis
remained undetected until St. 23/25 (head process stage). Those
observations agree with results reported by Mohun et al. (1980).

DISCUSSION

C^{14} amino acids and 2-D gel electrophoresis has been employed to
study the changes in protein synthesis of two species of amphibian
embryos. It can be concluded that even though the synthesis patterns
of most proteins remained unchanged during early development, a
number of significant changes do indeed occur. Figures 3 and 4
summarized the types of changes in protein synthesis observed in
axolotl and Xenopus embryos. Figure 7 compares the results of the
present study with data published previously by other investigators.

It is clear from both our results and previous work that no
change can be detected in the types of proteins synthesized during
oocyte maturation. That is surprising since an increase in the rate
of protein synthesis occurs around the time of germinal vesicle
breakdown during Xenopus oocyte maturation (Smith, 1975). The
increase in rate is not, however, reflected in the 2-D gel patterns
of either Xenopus or axolotl oocytes. Any changes that occur
apparently are below the limit of resolution of the 2-D gel methods.

Even though 2-D gel electrophoresis is highly sensitive for

| Stage of Development \ Study | MEULER and MALACINSKI | | BALLATINE *et al.* | BRAVO and KNOWLAND |
	Axolotl	Xenopus	Xenopus	Xenopus
Oocyte Maturation	+	+	+	ND
Fertilization	+	+	+ / −	+ / −
Cleavage	−	+	−	+
Blastula	+ / −	+ / −	+	+ / −
Gastrula	+ / −	+ / −	+	−
Neurula	+	+	+	ND
Tailbud	+	ND	+	ND
KEY: agree = (+) disagree = (−) no data = ND				

Figure 7: A comparison of the results of the present study with
  previously published reports (Ballantine et al., 1979; Bravo and
  Knowland, 1979).

separating and detecting polypeptides, it still has several
limitations. Only those proteins synthesized at relatively high
rates can be recognized (O'Farrell, 1975). Complexity studies
performed on poly(A)$^+$ mRNA populations in sea urchin gastrula have
demonstrated that transcripts from approximately 14,000 different
structural genes are present on gastrular polysomes (Galau et al.,
1974). Yet only approximately 750 sea urchin embryonic proteins can
be detected on 2-D gels. Two-dimensional gels of amphibian embryonic
proteins detect only approximately 270 proteins. Complexities of
amphibian oocyte RNA population species have been measured and found
to be quite similar to those of sea urchin oocytes (Galau et al.,
1976; Davidson and Hough, 1971). Therefore, the methods employed in
the present study detect only a very small proportion (perhaps less
than 10%) of the total oocyte mRNA translation product.

Approximately 40% of the total radioactivity applied to a gel is
found as discrete spots on a fluorograph. The remaining 60% of the
radioactivity either never enters the gel as discrete spots or is
present in the large streak, probably containing the histones and
other basic proteins, at the basic end of the gel. Because the pH
range of the isoelectric focusing gel is 5.0-7.2, the histones,
as well as other basic proteins, migrate to the basic end of the
first-dimension gel and are indicated by the large streak at one end
of the second-dimension gel. Therefore, less than 50% of the total
radioactivity representing the labeled proteins are present as
discrete spots on a fluorograph and are actually analyzed.

From those considerations it can be concluded that in this type

of analysis only a very small percentage of the total protein
population is being analyzed. The proteins seen on the gel therefore
represent those proteins synthesized at relatively high rates.

It was observed in these studies (Fig. 3 and 4) that in both
Xenopus and axolotl embryos many proteins synthesized during
oogenesis cease synthesis after fertilization and are subsequently
detected later in development. It would be interesting to understand
the types of regulatory events which control those changes in protein
synthesis at fertilization.

It might be possible to analyze the role either sperm
penetration or the sperm nucleus plays in regulating the pattern of
protein synthesis by using artificially activated eggs. Artificially
activated eggs undergo many of the normal responses to fertilization,
including a several-fold increase in the rate of protein synthesis
(Woodland, 1974). We have obtained preliminary results on
artificially activated Xenopus eggs which indicate that the changes
in protein synthesis occurring at fertilization (i.e., synthesis of
novel non-oogenic proteins and loss of many oogenic proteins) do
indeed occur. It therefore appears that neither the sperm nucleus
nor sperm entry affect the 2-D gel patterns found at fertilization.

The present report also demonstrated a difference between the
types of protein synthesis changes which occurred during gastrulation
in Xenopus versus the axolotl. In Xenopus embryos, a few novel
non-oogenic proteins were detected with the onset of gastrulation.
In the axolotl embryo, however, a more substantial change in the 2-D
gel pattern was found. Approximately 10% of the proteins detected in
axolotl pre-gastrula stages were no longer detected at gastrulation.
Furthermore, 10% of the proteins synthesized during gastrulation
represented the synthesis of novel non-oogenic proteins. These
proteins were subsequently found at all later stages of development.

The differences in gastrulation protein synthesis patterns
between the two species might be related to variations in the way the
two species undergo gastrulation. Gastrulation in Xenopus differs
from the axolotl mainly in the location of the prospective mesoderm
in early gastrula as well as the movements that bring the mesoderm to
its final position between the ectoderm and endoderm. In Xenopus
embryos the presumptive mesoderm is found in the deep layers beneath
the superficial cell layer of the early gastrula (Keller, 1976). In
axolotl embryos the presumptive mesoderm is found in the superficial
cells of the early gastrula (Smith and Malacinski, 1983).

It would be of interest to determine whether altering the
morphogenic movements of gastrulation affects protein synthesis.
Cell movements might be important in regulating those distinct
changes in protein synthesis observed at gastrulation in the
axolotl.

One way to approach this problem would be to examine the types of proteins synthesized in prospective germ layers that have been prevented from undergoing morphogenic movements. When presumptive ectoderm from blastula stage axolotl embryos is cultured in vitro, it rolls up into a ball. Preliminary results have indicated that the 2-D gel patterns of the axolotl ectoderm cultures display many of the changes that occur in whole control embryos which have undergone gastrulation. Many of these changes represent either the normal loss or gain of a protein and occur even though the cultured ectoderm cannot undergo gastrulation. The synthesis of those proteins apparently does not require the cell movements and tissue interactions that occur during gastrulation.

The 2-D gel patterns of axolotl ectoderm cultures, however, do not display all of the changes observed at gastrulation. A number of the proteins that begin detectable synthesis at gastrulation are not found in the 2-D gel patterns of cultured ectoderm. Furthermore, many of the pre-gastrula proteins that normally cease synthesis at gastrulation are still present. The synthesis of those proteins might be regulated by the cell movements and tissue interactions of gastrulation which are absent in ectoderm cultures.

It was also clear from both our results and previous work (Ballantine et al., 1979; Bravo and Knowland, 1979) (Fig. 3 and 4) that the incidence of novel non-oogenic protein synthesis increases after gastrulation in Xenopus and axolotl embryos. The majority of these proteins are found in all tissues of the embryo, while only a few AST or NAST specific proteins are detected. The apparent lack of tissue specific proteins in axolotl and Xenopus embryos is probably due to limitations of the resolution of the 2-D gel methods.

As mentioned before, 2-D gel electrophoresis detects a very small proportion of the potential mRNA translation products. Many of the proteins synthesized in specific tissues of the developing embryo are probably synthesized at low rates and would therefore not be detected. The AST and NAST specific proteins detected in the 2-D gel patterns of axolotl embryos are those proteins that have high rates of synthesis and are therefore easily detected by gel electrophoresis.

Nevertheless, novel non-oogenic proteins were detected after gastrulation for most stages of axolotl and Xenopus embryonic development. It would be interesting to know if altering the movements of gastrulation will affect the types of proteins synthesized after gastrulation.

Normal gastrulation movements can be altered by inducing exogastrulation. This can be accomplished by exposing an embryo to high salt or LiCl during gastrulation. The mesoderm and endoderm fail to invaginate but continue their gastrulation movements in the opposite direction, away from the ectoderm. The ectoderm becomes separated almost completely from the mesoderm and endoderm. In total exogastrulation, no neural structures are formed because the chordomesoderm is unable to interact with the presumptive neural ectoderm to form neural structures (Holtfreter, 1933).

Preliminary experiments on the effects of exogastrulation on the protein synthesis patterns of axolotl early neurula embryos revealed some surprising changes in the 2-D gel pattern. One percent LiCl or 200% Steinberg's solution-induced total exogastrulation led to the absence of the novel non-oogenic proteins that begin detectable synthesis at the onset of neurulation. But many proteins appear to be synthesized that were never before detected at any stage. These exogastrula specific proteins represent approximately 6% of the total proteins detected by fluorography.

Exogastrulae induced by 100% Steinberg's solution (pH 10.0) synthesized none of the exogastrula specific proteins detected in the 1.0% LiCl or 200% Steinberg's solution. They do, however, synthesize many of the novel non-oogenic proteins that begin detectable synthesis at neurulation. These proteins apparently do not require the inductive tissue interactions that result from gastrulation.

Some of the novel non-oogenic proteins found in the early neurula control embryos (including two AST specific proteins) were not detected in fluorographs of the 100% Steinberg's solution- (pH 10.0) induced total exogastrula. The synthesis of those proteins, especially the two AST specific proteins, might require the inductive tissue interactions resulting from gastrulation.

Finally, the discrepancy in the onset of α-actin synthesis between axolotl and Xenopus embryos has some interesting implications for the "differentiation" of the somitic mesoderm. Little is known about the events that result in cellular differentiation because it is difficult to precisely assess the time at which cellular differentiation begins.

One approach involves monitoring the synthesis of a differentiated cell product such as α-actin. Somitic mesoderm differentiates into a variety of tissues, including muscle cells. Alpha-actin is found specifically in muscle cells. In Xenopus embryos, α-actin begins detectable synthesis at gastrulation. It is difficult to imagine, however, that the somitic mesoderm in Xenopus embryos undergoes cell differentiation at a time when the primary germ layers are first being specified. In axolotl embryos, α-actin synthesis is initiated much later, at the head process stage (St. 23/25), in conjunction with the appearance of microfilaments in the somites. In this case the synthesis of a differentiated cell product may not serve as a marker for cell differentiation.

Another possible explanation might be that Xenopus embryos precociously synthesize α-actin in anticipation of the differentiation of the somitic mesoderm. Mohun et al. (1980) believed that because Xenopus embryos have a faster rate of development than most species of amphibians, it preloads its cells with α-actin in anticipation of muscle cell differentiation. In axolotl embryos, on the other hand, the developmental rate is much slower and the embryo does not need to anticipate the differentiation of the somites into muscle cells. Therefore, α-actin synthesis in the axolotl does not need to be initiated until that time when the somites actually begin differentiating into muscle. It would be interesting to determine whether other species of amphibians with developmental rates like Xenopus synthesize α-actin early in embryogenesis.

ACKNOWLEDGEMENTS

We are thankful to Gary Radice for technical assistance and for the gift of the rabbit muscle actin standard; Fran Briggs and Leah Dvorak for providing axolotl embryos; Sally Frost for providing moral support during the course of the investigation. This investigation was supported by NSF PCM 80-06343.

REFERENCES

Ballantine, J.E.M., Woodland, H.R., and Sturgess, E.A. (1979). Changes in protein synthesis during development of Xenopus laevis. J. Embryol. Exp. Morphol. 51:137-153.

Bordzilovskaya, N.P. and Dettlaff, T.A. (1979). Table of stages of the normal development of axolotl embryos and the prognostication of timing of successive developmental stages at various temperatures. Axolotl Colony Newsletter No. 7.

Brachet, J., Denis, H., and Devitry, F. (1964). The effects of Actinomycin D and puromycin on morphogenesis in amphibian eggs and Acetabularia mediterranea. Develop. Biol. 9:298-434.

Bravo, R. and Knowland, J. (1979). Class of proteins synthesized in oocytes, eggs, embryos, and differentiated tissues of Xenopus laevis. Differentiation 13:101-108.

Chamberlain, J. (1979). Fluorographic detection of radioactivity in polyacrylamide gels with the water-soluble fluor, sodium salicylate. Anal. Biochem. 98:132-135.

Cleveland, D.W., Fischer, S.G., Kirschner, M.W., and Laemmli, V.K. (1977). Peptide mapping by limited proteolysis in sodium dodecyl sulfate and analysis by gel electrophoresis. J. of Biol. Chem. 252:1102-1106.

Davidson, E.H. and Hough, B.R. (1971). Genetic information in oocyte RNA. J. Mol. Biol. 56:491-506.

Dumont, J.N. (1972). Oogenesis in Xenopus laevis (Daudin): I. Stages of oocyte development in laboratory maintained animals. J. Morph. 136:153-180.

Galau, G.A., Britten, R.J., and Davidson, E.H. (1974). A measurement of the sequence complexity of polysomal messenger RNA in sea urchin embryos. Cell 2:9-20.

Galau, G.A. (1976). Structural gene sets active in embryos and adult tissues of the sea urchin. Cell 7:487-505.

Gurdon, J.B. and Woodland, H.R. (1968). The cytoplasmic control of nuclear activity in animal development. Biol. Rev. 43:233-267.

Holtfreter, J. (1933). Die totale Exogastrulation, eine Selbstablosung des Ektoderms vom Entomesoderm. Entwicklung und funktionelles Verhalten nervenloser Organe. Arch. f. Entw'Mech. 129:669.

Keller, R.E. (1976). Vital dye mapping of the gastrula and neurula of Xenopus laevis: II. Prospective areas and morphogenetic movements of the deep layer. Develop. Biol. 51:118-137.

Mohun, T.J., Mohun, R., Tilly, R., and Slack, J. (1980). Cell commitment and gene expression in the axolotl embryo. Cell 22:9-15.

Nakahashi, T. and Yamana, K. (1974). Biochemical and cytological examination of the initiation of ribosomal-RNA synthesis during gastrulation of Xenopus laevis. Develop. Growth Differ. 18:329-338.

Newport, J. and Kirschner, M. (1982). A major developmental transition in early Xenopus embryos: II. Control of onset of transcription. Cell 30:687-696.

Niewkoop, P.D. and Faber, J. (1956). "Normal Table of Xenopus laevis (Daudin)." North-Holland, Amsterdam.

O'Farrell, P. (1975). High resolution 2-D electrophoresis of proteins. J. Biol. Chem. 1250:4007-4021.

Smith, J.C. and Malacinski, G.M. (1983). The origin of the mesoderm
   in Xenopus laevis, an anuran, and the axolotl, Ambystoma
   mexicanum, a urodele. Develop. Biol. (in press).

Smith, L.D. (1975). Molecular events during oocyte maturation. In
   "Biochemistry of Animal Development" Vol. 3, R. Webber (ed.).
   Academic Press, New York.

Sturgess, E.A., Ballantine, J.E.M., Woodland, H.R., Mohun, P.R.,
   Lane, C.D., and Dimitriadis, G.J. (1980). Actin synthesis
   during early development of Xenopus laevis. J. Embryol. Exp.
   Morph. 58:303-320.

Woodland, H.R. (1974). Changes in polysome content of developing
   Xenopus laevis embryos. Develop. Biol. 40:90-101.

CHANGES IN SYNTHESIS OF RNA AND PROTEIN DURING

REACTIVATION OF DELAYED IMPLANTING MOUSE BLASTOCYSTS

H.M. Weitlauf[*]

ABSTRACT

A series of experiments are reviewed that deal with various
aspects of RNA and protein synthesis during the developmental
diapause and reactivation associated with delayed implantation of
mouse embryos. Results of these experiments demonstrate that rates
of synthesis of RNA and protein are reduced during the quiescent
period associated with delayed implantation and increase with
reactivation of the embryos. The level of overall synthesis of RNA
was found to increase slowly but steadily over a 24 hr period,
reflecting primarily an increase in the rate of synthesis of
ribosomal RNA, while there appeared to be a marked but transient
burst in synthesis of mRNA within the first 3-4 hr. This increase in
synthesis of mRNA was found to be linked temporally to an α-amanitin
sensitive increase in protein synthesis. In addition, there appeared
to be an even earlier increase in protein synthesis that was
insensitive to α-amanitin. These results suggest that the overall
mechanism responsible for the embryonic diapause and reactivation of
delayed implanting mouse embryos has both transcriptional and
translational components. It is speculated that the primary response
to the maternal signal is at the level of translation of existing
mRNA and that the remainder of the metabolic changes and the
subsequent developmental arrest are a consequence of a reduction in
necessary proteins. Further experimental approaches to test this
hypothesis are suggested.

---

[*]From the Department of Anatomy, Texas Tech University Health
Center, Lubbock, TX 79430.

INTRODUCTION

The length of the pre-implantation phase of development may be extended in mice either by concurrent lactation or surgical removal of the mother's ovaries. In these situations, metabolic activity in the embryos is reduced and development is arrested at the blastocyst stage. The embryos are reactivated and resume development following removal of the suckling young or injection of estrogen into the mother. This phenomenon is generally referred to as delayed implantation (see McLaren, 1973; Renfree and Calaby, 1981, for references).

The molecular aspects of the mechanism that renders embryos metabolically and developmentally quiescent and then reactivates them in response to extrinsic signals from the mother are not known. Although it seems probable that regulation of protein synthesis plays a central role, this could be achieved either by changing the concentration of translatable mRNA or the rates at which ribosomes attach to or move along mRNA (i.e., initiation and elongation of peptide chains, Palmiter, 1972). Indeed, the results of several recent experiments in which synthetic activity in delayed implanting mouse embryos was examined during the process of reactivation seem to suggest that both translational and transcriptional controls are involved. The purpose of the present communication is to review briefly some of these findings and similar observations on normally developing embryos, with the hope that it will focus attention on the unanswered questions and stimulate further work in the area.

Overall Synthesis of RNA in Normal and Delayed Implanting Mouse Embryos

Early attempts to evaluate synthesis of RNA in pre-implantation mouse embryos were reported by Mintz (1964) and Monesi and Salfi (1967). In those studies, embryos of various ages were incubated in vitro with [^{3}H] uridine and incorporation of the tracer was evaluated by either autoradiographic or scintillation counting techniques. It was found that label was incorporated into RNA as early as the one-cell stage (i.e., after fertilization), labeling increased markedly in the four- to eight-cell stage, and then increased steadily through blastocyst formation. From those results it was generally inferred that synthesis of RNA in mouse embryos started early and increased markedly during cleavage and blastulation. However, one difficulty with such studies is that it cannot be assumed that changes in rates of incorporation of labeled uridine into RNA reflect equivalent changes in actual rates of RNA synthesis unless the specific activity of the intracellular pools of the immediate precursor to RNA are known and taken into account (Ellem and Gwatkin, 1968; Daentl and Epstein, 1971; Clegg and Piko, 1977; Warner, 1977). This caveat is generally applicable to tracer studies but it is particularly important in a system like the pre-implantation mouse embryo where uptake of uridine has been shown to change at the various developmental stages (Daentl and Epstein, 1971). Because of the small size of the embryos (pre-implantation mouse embryos contain from 1-190 cells [Kiessling and Weitlauf, 1979; Weitlauf, Kiessling, and Buschman, 1979] and have a dry mass of 25-35 ng [Lowenstein and Cohen, 1964; Hensleigh and Weitlauf, 1974]), it has not been practical to determine the specific activity of the UTP pools by a direct approach. Although attempts were made to circumvent this problem by saturating the intracellular pools with labeled uridine, that could not be achieved at all stages (Daentl and Epstein, 1971). For some time, then, it was generally recognized that total RNA synthesis was probably increasing during the period between days 2 and 3 of development, but that accurate measurement of the true rates of RNA synthesis at different stages of development would require the determination of the specific activities of the intracellular pools of UTP.

This difficulty was finally overcome by Clegg and Piko (1977). Those investigators utilized a sensitive assay for estimating the specific activity of labeled UPT in a small number of mouse embryos. The assay was based on the ability of RNA polymerase to synthesize the equal molar copolymer poly (rA-rU) from an artificial template [poly (dA-dT)], [^{3}H] UPT from the embryos, and exogenous [^{14}C] ATP. Briefly, a perchloric acid extract of the labeled embryos was obtained (i.e., containing the [^{3}H] UTP) and added to an assay mixture containing RNA polymerase, poly (dA-dT), and exogenous [^{14}C] ATP. The two keys to the assay are that: 1) ATP and UTP are incorporated in equal molar amounts and therefore the following relationship can be derived: ([^{3}H] cpm/[^{14}C] cpm)·k = (specific activity of [^{3}H]/specific activity of [^{14}C]), where k is a ratio of the counting efficiencies of the two isotopes; and 2) exogenous [^{14}C] ATP is added in sufficient excess so it is not diluted by endogenous ATP and thus its specific activity remains known. The product (i.e., poly [rA-rU]) was adsorbed onto glass fiber filter paper and washed extensively with cold TCA. The radioactivity (i.e., both [^{3}H] and [^{14}C]) in the acid precipitable product was estimated with a scintillation counter and the specific activity of the [^{3}H] UTP pool was calculated from the equation given above. With this method it was possible to demonstrate that following a 2 hr incubation in [^{3}H] uridine (i.e., at 43 Ci/mmole): 1) the rate of incorporation of [^{3}H] uridine into RNA was 4.4 x 10^{-6} and 3.3 x 10^{-2} pmole in one-celled embryos and blastocysts, respectively; 2) the specific activity of the intracellular pool of [^{3}H] UTP was 0.028 and 11.76 Ci/mmole; and 3) after correcting for the specific activities of UTP, the actual rates of synthesis of RNA were approximately 6.0 x 10^{-3} and 1.8 x 10^{-1} pmole uridine incorporated/embryo in 2 hr. Thus, in the face of an increase of approximately 10,000 fold in incorporation of label, the actual rate of RNA synthesis increased only 30 fold/embryo. Furthermore, if it is considered that the number of cells in the embryos also increases during this interval, the actual change in the rate of synthesis of RNA is on the order of only 1-2 fold per cell.

Several reports have appeared in the literature that deal with the synthesis of RNA by delayed implanting mouse embryos (Weitlauf, 1974, 1976, 1978; Chavez and Van Blerkom, 1979). In those experiments, blastocysts were generally recovered from normal or delayed implanting mice and incubated in vitro with [3H] uridine; rates of incorporation of label into acid insoluble RNA were determined by scintillation spectroscopy. With that approach it was shown that the rate of incorporation of labeled uridine by delayed implanting embryos was less than that by embryos that had been reactivated or were implanting normally. However, as has been noted above, the actual rates of RNA synthesis cannot be determined from the incorporation of labeled uridine unless the specific activity of the intracellular pools of UTP are known and taken into account.

Those measurements have recently been made by Weitlauf and Kiessling (1980). The preparation of animals used as donors of dormant and reactivated blastocysts in that experiment (and several other experiments from the same laboratory that are reviewed below) can be summarized as follows: Sexually mature virgin white Swiss mice were induced to ovulate with gonadotropins and placed with fertile males overnight. The finding of a vaginal plug on the following morning confirmed mating and that day was designated as day 1 of pregnancy. Animals to be used as donors of dormant embryos were ovariectomized bilaterally before noon on day 4 of pregnancy and injected with progesterone on days 7, 8, 9, and 10; those used to provide reactivated embryos were treated similarly but received estradiol-17β in addition to progesterone on days 9 and 10. For determination of overall rates of RNA synthesis, embryos were flushed from the uterus on day 10 and incubated in basal Eagle's medium (Eagle, 1955) with [3H] uridine (25 uCi/ml; 41.3 Ci/mmole) at 37°C in a moist atmosphere of 5% CO_2 in air for various intervals from 2 to 24 hr. The embryos were placed on glass fiber filter papers and washed extensively with cold TCA. The amount of radioactivity in acid insoluble RNA was determined with a scintillation counter. The amount of [3H] uridine incorporated into RNA with respect to time in vitro (i.e., for both active and dormant embryos) was reduced to a polynomial expression by regression analysis; the rates of incorporation of [3H] uridine were then calculated from the first derivatives of those expressions. Specific activities of the pools of [3H] UTP were determined by means of the double isotope technique of Clegg and Piko (1977). With the rates of incorporation of [3H] uridine into RNA and the specific activity of the [3H] UTP pools known, the overall rates of incorporation of uridine were calculated.

Initially the rate of incorporation of [3H] uridine in the active embryos was found to be approximately 2500 dpm/100 cells/hr, specific activity of the [3H] UTP was 16 pCi/fmole, and the actual rate of incorporation of uridine into RNA was therefore approximately 71 fmoles/100 cells/hr. Incorporation of [³H] uridine into the dormant embryos was approximately 1100 dpm/100 cells/hr, the specific activity of the [³H] UTP was 23 pCi/fmole, and the actual rate of uridine incorporation was 22 fmoles/100 cells/hr. Thus it appeared that the actual rate of overall RNA synthesis (or activity of RNA polymerase; Weitlauf and Kiessling, 1981) in delayed implanting embryos was about one-third of that in embryos that had been reactivated and were preparing to implant. Furthermore, incorporation of [³H] uridine by reactivated embryos was found to be linear for up to 24 hr in vitro (i.e., dy/dx = 770 cpm/100 cells/hr). By contrast, accumulation of label into the dormant embryos was initially low and increased with time in vitro (i.e., dy/dx = 14x + 309). When changes in the specific activity of the [³H] UTP were considered, the rate of overall uridine incorporation in active embryos was relatively constant (i.e., between 57 and 71 fmole/100 cells/hr; Fig. 1a), while the rate of uridine incorporation in delayed implanting embryos increased steadily as the embryos became reactivated in vitro (i.e., from 20 to 57 fmoles/100 cells/hr; Fig. 1b).

DNA Dependent RNA Polymerase Activity

Measurements of DNA dependent RNA polymerase activity in early mouse embryos have been undertaken with the general expectation that if amounts of enzyme are limiting for RNA synthesis in the embryos, then it should be possible to relate levels of activity at various developmental stages with the observed differences in synthesis of RNA (Siricusa, 1973; Siricusa and Vivarelli, 1975; Warner and Versteegh, 1974; Versteegh, Hearn, and Warner, 1975; Levey and Brinster, 1978). The investigators all used various modifications of the cell-free system devised by Roeder and Rutter (1970) to measure enzyme activity. Briefly, embryos were placed in small tubes, lysed, and added to an assay mixture containing DNA, Mn^{++}, Mg^{++}, ATP, CTP, GTP, [³H] UTP, and varying amounts of $[NH_4]_2SO_4$. The reaction was allowed to continue for some time at 37°C and the product was precipitated and washed extensively with cold TCA. In the earliest work with pre-implantation mouse embryos, Siricusa (1973) demonstrated that there was a generally increasing amount of RNA polymerase activity from the one-celled stage through the blastocyst stage. However, those measurements were made under conditions that did not favor either subtype of RNA polymerase (i.e., 45 mM $[NH_4]_2SO_4$), and therefore the studies were repeated with 125 mM or 25 mM $[NH_4]_2SO_4$ (high and low salt concentrations maximize for types II and [I + III], respectively). It was found that the amount of activity (i.e., at both salt concentrations) increased with development. However, based on estimates of cell numbers, it appeared that the amounts of both types of RNA polymerase activity per cell decreased after the eight-cell stage. Data from Levey and Brinster (1978) confirm the observation that amounts of enzymatic

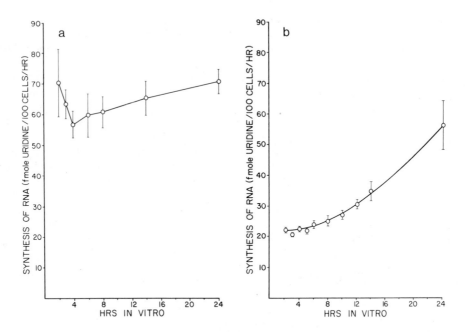

Figure 1: Overall rates of RNA synthesis in a) reactivated,
b) dormant mouse embryos incubated in vitro for 2-24 hr. Values
are means of 5-10 determinations ± SEM (modified by Weitlauf and
Kiessling, 1981).

activity increase with development from eight-cell to blastocyst; the
change may be as much as 12 amole/embryo/min when measured in high
salt and 9 amole/embryo/min when measured in low salt. Warner (1977)
has argued on the basis of "better" cell counts that, rather than
decreasing, the amount of activity per cell remains pretty much
constant after the eight-cell stage. However, it is difficult to
critically evaluate these claims without information regarding the
actual number of cells in embryos under the conditions employed in
the various laboratories.

In the next generation of RNA polymerase studies, combinations
of high and low salt concentrations were used along with high and low
concentrations of α-amanitin (a low concentration of α-amanitin
blocks type II RNA polymerase, a high concentration blocks both types
II and III). This was done in order to determine whether the
relative proportions of the various types of enzyme activity change
at different stages of development in a similar way as the types of

RNA being synthesized (Versteegh, Hearn, and Warner, 1975; Warner and Hearn, 1977; Levey and Brinster, 1978). From these kinds of studies it was suggested that the total amount of type II RNA polymerase in the eight-cell embryos was greater than the amount of a combination of types I and III, but that this situation was reversed at the blastocyst stage. In the final analysis, however, it was concluded by Siricusa and Vivarelli (1975) that there was no obvious relationship between amounts of enzyme activity and synthesis of major classes of RNA in pre-implantation mouse embryos. Warner (1977; Warner and Hearn, 1977) also concluded that changes in the proportions of type I, II, and III activities at eight-cell and blastocyst stages were not reflected by differences in the ratio of ribosomal, messenger, and transfer RNAs that are synthesized at these stages.

Measurements of DNA-dependent RNA polymerase activity have also been made in embryos from delayed implanting mice and on embryos reactivated after a period of quiescence (Weitlauf, 1981). Dormant or reactivated blastocysts were recovered and assayed as described above. The activity necessary to incorporate 1 fmole of UTP into RNA in 20 min was defined as 1 enzyme unit (EU). The reaction was shown to be dependent on exogenous DNA and the amount of product formed was proportional to the number of embryos assayed. It was found that when reactivated embryos were assayed in high salt, they contained 0.98 EU/100 cells (or incorporated 87 amole UTP/embryo/min), while the dormant embryos contained 0.58 EU/100 cells (or incorporated 43 amole UTP/embryo/min). When embryos were assayed in high salt with a low concentration of α-amanitin, or in low salt, there was a proportionally greater decrease in activity in reactivated embryos than in dormant embryos. These results seemed to suggest that there was a relatively greater proportion of type II RNA polymerase in the active embryos than in the dormant embryos. Furthermore, mixing experiments suggested the presence of an inhibitor of RNA polymerase activity in the delayed implanting embryos. However, it should be noted that a potential problem in interpreting the results of experiments using this cell-free assay for measuring RNA polymerase activity in vitro is that the amount of enzymatic activity measured tends to be less than 5-10% of that measured in intact cells (Roeder and Rutter, 1970). It is not clear, in general, whether the activities of the various RNA polymerases are reduced proportionately or there is a selective reduction of activities of particular forms of the enzymes, and this is certainly not known for the mouse blastocyst. A further complication is that although changes in salt concentrations will favor activity of particular forms of the enzyme, neither this nor the effects of α-amanitin can be considered absolutely selective (Levey and Brinster, 1978; Weitlauf, 1981). Thus it is difficult to interpret the results of such studies by themselves, and beyond making general observations about relative amounts of activity at different developmental stages, little can be said.

Major Classes of RNA

    From the results of the earliest autoradiograph studies, it was
inferred that synthesis of nuclear RNA (presumed messenger RNA)
occurs throughout all stages--from fertilization to the
blastocyst--and that synthesis of nucleolar RNA (i.e., ribosomal RNA)
and its movement to the cytoplasm probably started at about the
four-cell stage (Mintz, 1964). Direct evidence in support of these
suppositions was subsequently provided by chromatography with
methylated bovine serum albuminkieselguhr (MAK) (Ellem and Gwatkin,
1968). In that study it was shown that the dominant forms of RNA
synthesized from the four-cell stage onward were 18S and 28S
ribosomal RNA and 4S transfer RNA, although synthesis of high and low
molecular weight heterogenous DNA-like RNA (presumptive mRNA) was
also detected. These observations have subsequently been confirmed
and refined by others (Woodland and Graham, 1969; Piko, 1970;
Knowland and Graham, 1972; Piko, 1975; Levey, Stull, and Brinster,
1978; Kidder and Pederson, 1982; see Schultz and Tucker, 1977, for a
review). More recently it has been shown that polyadenylated RNA
(i.e., mRNA) and 18S and 28S ribosomal RNA are synthesized as early
as the two-cell stage (Clegg and Piko, 1982).

    Indirect evidence that transcription of RNA occurs and is
important in early development is provided by the finding that both
Actinomycin D and α-amanitin block development, maybe as early as the
one-cell stage, but certainly after the two-cell stage (see Warner,
1977; Warner and Versteegh, 1974; Globus, Calarco, and Epstein, 1973;
Braude, 1979a; and Johnson, Handyside, and Braude, 1977; for
reviews). Further evidence that mRNA is transcribed in early mouse
embryos comes from the demonstration that paternal isozyme variants
show up in the cleavage stage embryos, paternal β-glucoronidase as
early as four-cell (Wudl and Chapman, 1976), and the paternal allele
for glucose phosphate isomerase and H-Y antigen as early as the
eight-cell stage (Brinster, 1973; Krco and Goldberg, 1976).

    Measurements of changes in rates of synthesis of
nonpolyadenylated RNA and polyadenylated RNA during reactivation of
delayed implanting embryos have recently been made (Weitlauf, 1982).
Animals were prepared, and dormant and fully activated blastocysts
were recovered as described above. In addition, embryos were
recovered at various intervals after the injection of estradiol-17β
(i.e., between 0 and 24 hr). Thus, blastocysts were collected at
intermediate points in reactivation as well as those that were at the
beginning or end of the process. A double label approach was used;
fully active embryos were incubated in vitro with 0.6 µM (^{14}C]
uridine (522 mCi/mmole) to provide a reference (control) level of RNA
synthesis; embryos undergoing reactivation were incubated with 0.6 µM
[^{3}H] uridine (40 Ci/mmole) for comparison. Known numbers of fully
active embryos (i.e., those labeled with [^{14}C]) and those undergoing
reactivation (i.e., labeled with [^{3}H]) were mixed together and the

differentially labeled RNA was extracted by the method of Chirgwin et al. (1979). Nonpolyadenylated RNA extracted by this method is largely rRNA; polyadenylated RNA is mRNA and possibly its polyadenylated precursor. Separation of rRNA and mRNA was accomplished by means of affinity chromatography with oligo(dT) cellulose (Aviv and Leder, 1972). The specific activities of the pools of intracellular UTP were determined by the method of Clegg and Piko (1977) as outlined above, and rates of incorporation of labeled uridine were corrected appropriately. It was assumed that the rate of accumulation of labeled RNA was proportional to the rate of synthesis of RNA; overall synthesis in the embryos undergoing activation was expressed as a proportion of that in fully active embryos.

The rate of synthesis of rRNA by dormant embryos was initially 35% of the control level; it increased slowly but steadily throughout the 24 hr as reactivation occurred (Fig. 2a). By contrast, the synthesis of mRNA was initially 46% of that in fully active embryos and it increased rapidly (i.e., within 3-4 hr) to 74% of the control value (Fig. 2b). Interestingly, this burst in synthesis of mRNA was transient, subsequently the rate of synthesis of mRNA followed that of rRNA closely (compare Fig. 2a and 2b).

Protein Synthesis

In mouse embryos the rate of incorporation of labeled amino acids increases appreciably between the eight-cell stage and blastocyst formation (Monesi and Salfi, 1967). Although the difficulties in relating changes in the incorporation of labeled amino acids and rates of protein synthesis are similar to those discussed above with respect to labeled uridine and RNA, the studies of Epstein and Smith (1973) confirm that actual rates of protein synthesis increase in the early embryos. With two-dimensional electrophoresis, it has been shown very nicely that there are several hundred polypeptides synthesized during this early embryonic period (Van Blerkom and Brockaway, 1975a; Braude, 1979b; Van Blerkom, Chavez, and Bell, 1979). Furthermore, the greatest qualitative change appears to occur between the one- and eight-cell stage, and although most of the proteins observed are common for both the ICM and trophoblast, some appear to be unique for each cell population (Epstein and Smith, 1973; and Howe and Solter, 1979).

It has been known for some time that the level of protein synthesis in delayed implanting mouse embryos is lower than it is in either normally implanting blastocysts or those that are reactivated after a period of developmental quiescence (Greenwald and Everett, 1959; Weitlauf and Greenwald, 1968; Van Blerkom and Brockaway, 1975b). More recently, embryos have been examined during the process of reactivation to determine when the change in incorporation of amino acids occurs (Weitlauf, 1973). Active (i.e., normally

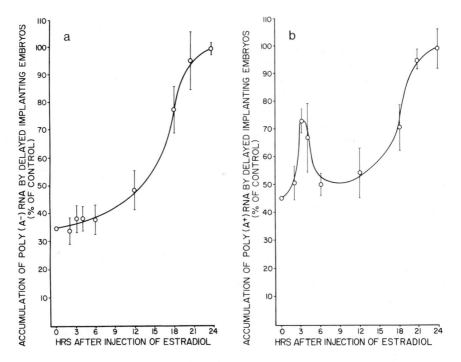

<u>Figure 2</u>: Accumulation of a) poly(A-) RNA and b) poly(A+) RNA by
mouse embryos as they leave the dormant phase associated with
delayed implantation. Data were corrected for numbers of cells and
are expressed as the rate of accumulation of RNA by delayed
implanting embryos divided by that in implanting embryos (x 100)
(i.e., percentage of control). Values are means of 5-10
determinations (± SEM) (modified from Weitlauf, 1982).

implanting) and dormant blastocysts were incubated <u>in vitro</u> for
varying periods of time (i.e., pre-incubation) and then pulsed with a
mixture of labeled amino acids ([^{14}C] leucine, lysine, arginine,
valine; 250-312 mCi/mmole). It was found that the rate of amino acid
incorporation by the active embryos was essentially constant
regardless of the length of time <u>in vitro</u>. By contrast, with delayed
implanting blastocysts it was found that the length of pre-incubation
did influence the rate of amino acid incorporation; thus, up to 4 hr,
the rate was found to be less than in normal embryos, but from 4-6 hr
it approached the normal level. Furthermore, actinomycin D (0.1
ug/ml) did not appear to interfere with this process (Weitlauf,
1973b).

In a related experiment, active and dormant blastocysts were

again incubated in vitro with labeled amino acids (Weitlauf, 1974).
In this case the length of incubation varied from 0.5 to 8 hr. Total
radioactivity accumulated in acid insoluble protein was determined as
before and the data expressed as a polynomial after regression
analysis; the slope of the resulting expressions indicating the rate
of incorporation of isotope. The rate of accumulation of
radioactivity by normal embryos was constant. By contrast, it was
relatively low in delayed implanting embryos when they were initially
removed from the uterus but increased to near the normal level after
a few hours in vitro. Actinomycin D (10 ug/ml) had little effect on
the rate of incorporation by active blastocysts; however, the
expected increase by delayed implanting embryos was blocked
completely. In cultured HeLa cells, Actinomycin D at a concentration
of 0.1 ug/ml is reported to block synthesis of rRNA but not mRNA; at
a concentration of 10 ug/ml it stops synthesis of both classes of RNA
(Perry, 1963).

This result, in conjunction with the observation reviewed above
of a transient burst in synthesis of mRNA early in reactivation,
suggested the possibility that control of transcription is involved
in the mechanism that regulates this process. Therefore, experiments
were undertaken to determine whether the inhibition of synthesis of
mRNA with α-amanitin would have an effect on the resumption of
protein synthesis during reactivation. Dormant embryos were obtained
as outlined above and pre-incubated for 8 hr in medium containing
various concentrations of α-amanitin (i.e., 0-50 ug/ml); embryos were
then pulsed with labeled amino acids (0.5 hr; 20 uCi/ml; [^{3}H] L-amino
acid mixture from New England Nuclear). The embryos were rinsed in
several ml of nonlabeled medium, placed on nitrocellulose filters,
and washed extensively with hot TCA. Radioactivity in the acid
insoluble protein was estimated with a scintillation spectrometer.
In addition, some embryos were incubated with labeled amino acids
(i.e., for 0.5 hr) immediately after being recovered from the uterus
to provide an estimate of the level of amino acid incorporation
during delayed implantation. Although the results are preliminary,
it was found, as expected, that the rate of incorporation of labeled
amino acids by embryos immediately after their recovery from the
uterus was low compared to that by embryos that had been
pre-incubated in vitro for 8 hr (i.e., 68 and 270 cpm/embryo/0.5 hr,
respectively). Furthermore, it was found that α-amanitin reduced the
rate of amino acid incorporation in a dose dependent manner (Fig. 3),
although it was never reduced to the level observed in quiescent
embryos.

In a second experiment, dormant embryos were recovered as
described above and immediately incubated with α-amanitin (30 ug/ml;
0.5 hr); the embryos were removed from the inhibitor, washed, and
placed in fresh medium. Some embryos were subsequently recovered at
each of the intervals shown in Fig. 4 and pulsed with labeled amino
acids (for 0.5 hr). Radioactivity incorporated into protein was

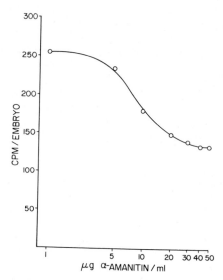

<u>Figure 3</u>: Incorporation of labeled amino acids by delayed implanting
    embryos undergoing reactivation <u>in vitro</u>. Embryos were
    pre-incubated for 8 hr in medium with varying concentrations of
    α-amanitin (i.e., 0-50 µg/ml) and pulsed with [^{3}H] amino acids for
    0.5 hr.

estimated as outlined above. The expected increase in the rate of
incorporation of labeled amino acid was found to occur in two
increments. The initial increase occurred within 0.5 hr of the time
embryos were removed from the uterus (i.e., a change from 68 to 200
cpm/embryo/0.5 hr; Fig. 4). The second increase occurred after the
embryos had been <u>in vitro</u> for 4-5 hr (i.e., a change from 200-300
cpm/embryo/0.5 hr; Fig. 4). Although the initial increase in the
rate of amino acid incorporation was not affected by treatment with

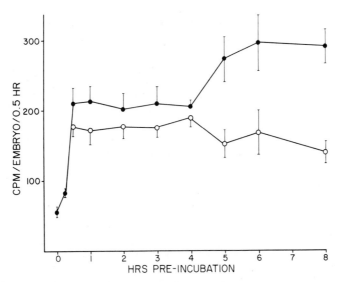

Figure 4: Incorporation of labeled amino acids by dormant
blastocysts that were exposed to α-amanitin (30 µg/ml) for 0.5 hr
immediately after being recovered from the uterus (○); or exposed
to medium without inhibitor (●). The embryos were then placed in
fresh medium, recovered at various intervals, and pulsed with [³H]
amino acids (mean ± SEM, 5-10 determinations).

α-amanitin, the second increase did not occur in embryos that had
been exposed to the inhibitor. Thus, from these preliminary
observations, it appears that during reactivation some changes in
protein synthesis require synthesis of new mRNA while others do not.
Many examples of control of protein synthesis have been reported at
both the level of transcription and translation in other tissues (see
Palmiter, 1972; Fan and Penman, 1970; Nielsen and McConkey, 1980; for
references)

CONCLUDING COMMENTS

It seems clear from the foregoing that pre-implantation mouse embryos have the capability of synthesizing protein and the major classes of RNA as early as the one- or two-cell stage and that their synthetic capacity, in general, increases through blastocyst formation. Furthermore, because a few proteins appear or disappear, and because α-amanitin will stop development at almost any stage, it seems that both transcription and translation of new mRNA are occurring more or less throughout early development. Although these observations and the demonstration that there are quantitative differences in synthetic activity by dormant and reactivated delayed implanting embryos are interesting and important for an understanding of early development, they do not provide much insight into putative regulatory mechanisms that might be utilized to cause the embryonic diapause associated with delayed implantation.

On the other hand, the observation that the overall increase in protein synthesis that occurs during reactivation of dormant embryos has two components does appear to provide clues about control mechanisms and indicates a direction for further experimentation. Because the early increase in protein synthesis is insensitive to α-amanitin, it is probable that new mRNA is not essential for its occurrence. By contrast, the second phase was found to be temporally related to a burst in synthesis of mRNA and to be inhibited by α-amanitin. It appears, therefore, that the overall mechanism responsible for controlling protein synthesis during reactivation of delayed implanting embryos involves regulation at the levels of both translation and transcription.

Because of the temporal relationships, it is tempting to suggest that the primary control in response to maternal signals is exerted at the level of translation and that the remaining metabolic changes and developmental arrest occur as a consequence. Among the questions to be answered next are: 1) Is the translational step controlled by regulating initiation or elongation of peptide chains? 2) Does the increase in transcription depend on the earlier change in translation? 3) Can changes in specific proteins be related to either phase of reactivation? It is to be hoped that with the answers to these questions the basic mechanism responsible for the embryonic diapause associated with delayed implantation will become clear.

ACKNOWLEDGEMENT

Support by National Institutes of Health is gratefully acknowledged (HD 08496).

REFERENCES

Aviv, H. and Leder, R. (1972). Purification of biologically active globin messenger RNA by chromatography on oligothymidylic acid-cellulose. Proc. Natl. Acad. Sci. USA 69:1408-1412.

Braude, P.R. (1979a). Time-dependent effects of α-amanitin on blastocyst formation in the mouse. J. Embryol. Exp. Morph. 52:193-202.

Braude, P.R. (1979b). Control of protein synthesis during blastocyst formation in the mouse. Develop. Biol. 68:440-452.

Brinster, R.L. (1973). Paternal glucose phosphate isomerase activity in three-day mouse embryos. Biochem. Genet. 9:187-191.

Chavez, D.J. and Van Blerkom, J. (1979). Persistance of embryonic RNA synthesis during facultative delayed implantation in the mouse. Develop. Biol. 70:39-49.

Chirgwin, J.M., Przybla, A.E., MacDonald, R.J., and Rutter, W.J. (1979). Isolation of biologically active ribonucleic acid from sources enriched in ribonuclease. Biochemistry 18:5294-5299.

Clegg, K.B. and Piko, L. (1977). Size and specific activity of the UTP pool and overall rates of RNA synthesis in early mouse embryos. Develop. Biol. 58:76-95.

Clegg, K.B. and Piko, L. (1982). RNA synthesis and cytoplasmic polyadenylation in the one-cell mouse embryo. Nature 295, Number 5847, pp. 342-344.

Daentl, D.L. and Epstein, D.J. (1971). Developmental interrelationships of uridine uptake, nucleotide formation, and incorporation into RNA by early mammalian embryos. Develop. Biol. 31:316-322.

Eagle, H. (1955). Nutrition needs of mammalian cells in tissue culture. Science 122:501-504.

Ellem, K.A.O. and Gwatkin, R.B.L. (1968). Patterns of nucleic acid synthesis in the early mouse embryo. Develop. Biol. 18:311-330.

Epstein, C. and Smith, S. (1973). Amino acid uptake and protein synthesis in pre-implantation mouse embryos. Develop. Biol. 33:171-184.

Fan, H. and Penman, S. (1970). Regulation of protein synthesis in mammalian cells. II. Inhibition of protein synthesis at the level of initiation during mitosis. J. Mol. Biol. 50:655-670.

Globus, M.S., Calarco, P.G., and Epstein, C.J. (1973). The effects of inhibitors of RNA synthesis (α-amanitin and Actinomycin D) on pre-implantation mouse embryogenesis. J. Exp. Zool. 186:207-216.

Greenwald, G.S. and Everett, N.B. (1959). The incorporation of S35 methionine by the uterus and ova of the mouse. Anat. Rec. 134:171-184.

Hensleigh, H.C. and Weitlauf, H.M. (1974). Effect of delayed implantation on dry weight and lipid content of mouse blastocysts. Biol. Reprod. 10:315-320.

Howe, C.C. and Solter, D. (1979). Cytoplasmic and nuclear protein synthesis in pre-implantation mouse embryos. J. Embryol. Exp. Morph. 52:209-225.

Johnson, M.H., Handyside, A.H., and Braude, P.R. (1977). Control mechanisms in early mammalian development. In "Development in Mammals," M. Johnson (ed.), Vol. 2. North Holland, New York.

Kidder, G.M. and Pedersen, R.A. (1982). Turnover of embryonic messenger RNA in pre-implantation mouse embryos. J. Embryol. Exp. Morph. 67:37-49.

Kiessling, A.A. and Weitlauf, H.M. (1979). DNA polymerase activity in pre-implantation mouse embryos. J. Exp. Zool. 208:347-354.

Knowland, J. and Graham, C.F. (1972). RNA synthesis at the two-cell stage of mouse development. J. Embryol. Exp. Morph. 27:167-176.

Krco, C.J. and Goldberg, E.H. (1976). H-Y (male) antigen: detection on eight-cell mouse embryos. Science 193:1134-1135.

Levey, I.L. and Brinster, R.L. (1978). Effects of α-amanitin on RNA synthesis by mouse embryos in culture. J. Exp. Zool. 203:351-360.

Levey, I.L., Stull, G.B., and Brinster, R.L. (1978). Poly(A) and synthesis of polyadenylated RNA in the pre-implantation mouse embryo. Develop. Biol. 64:140-148.

Lowenstein, J.E. and Cohen, A.I. (1964). Dry mass, lipid content, and protein content of the intact and zona-free mouse ovum. J. Embryol. Exp. Morph. 12:113-121.

McLaren, A. (1973). Blastocyst activation. In "The Regulation of Mammalian Reproduction," S.J. Segal, R. Crozier, P.A. Corfman, and P.G. Condliffe (eds.), pp. 321-328. Charles C. Thomas, Springfield, Ill.

Mintz, B. (1964). Synthetic processes and early development in the mammalian egg. J. Exp. Zool. 157:85-100.

Monesi, V. and Salfi, V. (1967). Macromolecular synthesis during early development in the mouse embryo. Exp. Cell Res. 46:632-635.

Nielsen, P.J. and McConkey, E.H. (1980). Evidence for control of protein synthesis in HeLa cells via the elongation rate. J. Cell Physiol. 104:269-281.

Palmiter, R.D. (1972). Regulation of protein synthesis in chick oviduct. II. Modulation of polypeptide elongation and initiation rates by estrogen and progesterone. J. Biol. Chem. 247:6770-6780.

Perry, R.P. (1963). Selective effects of Actinomycin D on the intracellular distribution of RNA synthesis in tissue culture cells. Exp. Cell Res. 29:400-406.

Piko, L. (1970). Synthesis of macromolecules in early mouse embryos cultured in vitro: RNA, DNA, and A polysaccharide component. Develop. Biol. 21:257-279.

Piko, L. (1975). Expression of mitochondrial and nuclear genes during early development. In "The Early Development of Mammals," M. Balls and A.E. Wilds (eds.), pp. 167-178. Cambridge University Press, Cambridge.

Renfree, M.B. and Calaby, J.H. (1981). Background to delayed implantation and embryonic diapause. In "Embryonic Diapause in Mammals," A.P.F. Flint, M.B. Renfree, and B.J. Weir (eds.). J. Reprod. Fert. Suppl. 29:1-9.

Roeder, R.G. and Rutter, W.J. (1970). Multiple ribonucleic acid polymerases and ribonucleic acid synthesis during sea urchin development. Biochemistry 9:2543-2553.

Schultz, G.A. and Tucker, E.B. (1977). Protein synthesis and gene expression in pre-implantation rabbit embryos. In "Development in Mammals," M. Johnson (ed.), Vol. 1. North Holland, New York.

Siracusa, G. (1973). RNA polymerase during early development in mouse embryo. Exptl. Cell Res. 78:460-462.

Siracusa, G. and Vivarelli, E. (1975). Low-salt and high-salt RNA polymerase activity during pre-implantation development in the mouse. J. Reprod. Fert. 43:567-569.

Van Blerkom, J. and Brockaway, G.O. (1975a). Qualitative patterns of protein synthesis in the pre-implantation mouse embryo. I. Normal pregnancy. Develop. Biol. 44:148-157.

Van Blerkom, J. and Brockaway, G.O. (1975b). Qualitative patterns of protein synthesis in the pre-implantation mouse embryo. II. During release from facultative delayed implantation. Develop. Biol. 46:446-451.

Van Blerkom, J. Chavez, P.J., and Bell, H. (1979). Molecular and cellular aspects of facultative delayed implantation in the mouse. In "Maternal Recognition of Pregnancy," pp. 141-163. Ciba Foundation Symposium 65 (New Series).

Versteegh, L.R., Hearn, T.F., and Warner, C.M. (1975). Variations in the amounts of RNA polymerase forms I, II, and III during pre-implantation development in the mouse. Develop. Biol. 46:430-435.

Warner, C.M. (1977). RNA polymerase activity in pre-implantation mammalian embryos. In "Development in Mammals," M. Johnson (ed.), Vol. 1. North Holland, New York.

Warner, C.M. and Hearn, T.F. (1977). The effect of α-amanitin on nucleic acid synthesis in pre-implantation mouse embryos. Differentiation 7:89-97.

Warner, C.M. and Versteegh, L.R. (1974). In vivo and in vitro effect of α-amanitin on pre-implantation mouse embryo RNA polymerase. Nature 248:678-680.

Weitlauf, H.M. (1973a). In vitro uptake and incorporation of amino acids by blastocysts from intact and ovariectomized mice. J. Exp. Zool. 183:303-308.

Weitlauf, H.M. (1973b). Effect of Actinomycin D on activation of delayed implanting mouse blastocysts in vitro. Anat. Rec. 175:466A.

Weitlauf, H.M. (1974). Effect of Actinomycin D on protein synthesis by delayed implanting mouse embryos in vitro. J. Exp. Zool. 189:197-202.

Weitlauf, H.M. (1976). Effect of uterine flushings on RNA synthesis by 'implanting' and 'delayed implanting' mouse blastocysts in vitro. Biol. Reprod. 14:566-571.

Weitlauf, H.M. (1978). Factors in mouse uterine fluid that inhibit
    the incorporation of [^{3}H] uridine in vitro. J. Reprod. Fert.
    52:321-325.

Weitlauf, H.M. (1981). In vitro measurements of RNA polymerase
    activity in implanting and delayed-implanting mouse embryos.
    Develop. Biol. 83:178-182.

Weitlauf, H.M. (1982). A comparison of the rates of accumulation of
    nonpolyadenylated and polyadenylated RNA in normal and delayed
    implanting mouse embryos. Develop. Biol. 93:266-271.

Weitlauf, H.M. and Greenwald, G.S. (1968). Influence of estrogen
    and progesterone on the incorporation of ^{35}S methionine by
    blastocysts in ovariectomized mice. J. Exp. Zool. 169:463-470.

Weitlauf, H.M. and Kiessling, A.A. (1980). Comparison of overall
    rates of RNA synthesis in implanting and delayed implanting
    mouse blastocysts in vitro. Develop. Biol. 77:116-129.

Weitlauf, H.M. and Kiessling, A.A. (1981). Activity of RNA and DNA
    polymerases in delayed-implanting mouse embryos. In "Cellular
    and Molecular Aspects of Implantation," S.R. Glasser and D.W.
    Bullock (eds.), pp. 125-136. Plenum Publishing Company, New
    York.

Weitlauf, H.M., Kiessling, A., and Buschman, R. (1979). Comparison
    of DNA polymerase activity and cell division in normal and
    delayed-implanting mouse embryos. J. Exp. Zool. 209:467-472.

Woodland, H.R. and Graham, C.F. (1969). RNA synthesis during early
    development of the mouse. Nature (London) 221:327-332.

Wundl, L. and Chapman, V. (1976). The expression of β-glucoronidase
    during pre-implantation development of mouse embryos. Develop.
    Biol. 48:104-10>.

CONTRIBUTORS

Abbott, Michael K.
    Indiana University, Bloomington, IN 47405

Angerer, Lynne M.
    University of Rochester, Rochester, NY 14618

Angerer, Robert C.
    University of Rochester, Rochester, NY 14618

Anstrom, John A.
    Indiana University, Bloomington, IN 47405

Bedard, Pierre-Andre
    McGill University, Montreal, Quebec, Canada

Boyer, Paul David
    University of California, Los Angeles, CA 90024

Brandhorst, Bruce P.
    McGill University, Montreal, Quebec, Canada

Brodeur, Richard D.
    University of Texas, Austin, TX 78712

Bruskin, Arthur M.
    Indiana University, Bloomington, IN 47405

Carpenter, Clifford D.
    Indiana University, Bloomington, IN 47405

Colot, Hildur V.
    Brandeis University, Waltham, MA 02254

DeLeon, Donna V.
    University of Rochester, Rochester, NY 14618

Eldon, Elizabeth D.
        Indiana University, Bloomington, IN 47405

Graham, Melissa L.
        University of California, Los Angeles, CA 90024

Hughes, Kathleen J.
        University of Rochester, Rochester, NY 14618

Hursh, Deborah A.
        Indiana University, Bloomington, IN 47405

Hyman, Linda E.
        Brandeis University, Waltham, MA 02254

Jeffery, William R.
        University of Texas, Austin, TX 78712

Kaufman, Thomas C.
        Indiana University, Bloomington, IN 47405

Klein, William H.
        Indiana University, Bloomington, IN 47405

Konrad, Kenneth D.
        Case Western Reserve University, Cleveland, OH 44106

Leaf, David S.
        Indiana University, Bloomington, IN 47405

Lee, Inyong
        University of California, Los Angeles, CA 90024

Lengyel, Judith A.
        University of California, Los Angeles, CA 90024

Lynn, David A.
        University of Rochester, Rochester, NY 14618

Mahowald, Anthony P.
        Case Western Reserve University, Cleveland, OH 44106

Malacinski, George M.
        Indiana University, Bloomington, IN 47405

Meier, Stephen
        University of Texas, Austin, TX 78712

Meuler, Debbie Crise
        Indiana University, Bloomington, IN 47405

Raff, Rudolf A.
    Indiana University, Bloomington, IN 47405

Roark, Margaret
    University of California, Los Angeles, CA 90024

Rosbas, Michael
    Brandeis University, Waltham, MA 02254

Rosenthal, Eric T.
    Harvard Medical School, Boston, MA 02115

Ruderman, Joan V.
    Harvard Medical School, Boston, MA 02115

Salas, Fidel
    University of California, Los Angeles, CA 90024

Showman, Richard M.
    Indiana University, Bloomington, IN 47405

Spain, Lisa M.
    Indiana University, Bloomington, IN 47405

Strecker, Teresa R.
    University of California, Los Angeles, CA 90024

Strome, Susan
    University of Colorado, Boulder, CO 80309

Tansey, Terese
    Harvard Medical School, Boston, MA 02115

Thomas, Steven R.
    University of California, Los Angeles, CA 90024

Tomlinson, Craig R.
    University of Texas, Austin, TX 78712

Tufaro, Frank
    McGill University, Montreal, Quebec, Canada

Tyner, Angela L.
    Indiana University, Bloomington, IN 47405

Underwood, Eileen M.
    University of California, Los Angeles, CA 90024

Wells, Dan E.
    Indiana University, Bloomington, IN 47405

Weitlauf, H.M.
    Texas Tech University, Lubbock, TX   79430

Wood, William B.
    University of Colorado, Boulder, CO   80309

PHOTOS OF PARTICIPANTS

TOP ROW - Left to right: Brandhorst, Fabian, Morton and Klein.

BOTTOM ROW - Left: Kaufman
             Right: Group Photo: Standing - Malacinski, Strome,
                    Jeffrey, Lengyel, Mahowald, Ruderman, Raff,
                    Hyman and Brandhorst.
                    Kneeling - Weitlauf, Kaufman, Costantini, Klein
                    and Angerer.

313

Ohio River   Strome

TOP ROW: Mahowald, Kaufman and Raff

BOTTOM ROW: Weilauf, Processed Spissula and Lengyel.

INDEX

315